MOBILE CLOUD COMPUTING

Models, Implementation, and Security

MOBILE CLOUD COMPUTING

Models, Implementation, and Security

Meikang Qiu

Pace University
New York City, New York, USA

Keke Gai

Pace University
New York City, New York, USA

CRC Press
Taylor & Francis Group
Boca Raton London New York

CRC Press is an imprint of the
Taylor & Francis Group, an **informa** business

A CHAPMAN & HALL BOOK

CRC Press
Taylor & Francis Group
6000 Broken Sound Parkway NW, Suite 300
Boca Raton, FL 33487-2742

© 2017 by Taylor & Francis Group, LLC
CRC Press is an imprint of Taylor & Francis Group, an Informa business

No claim to original U.S. Government works

Printed on acid-free paper
Version Date: 20170413

International Standard Book Number-13: 978-1-4987-9603-3 (Hardback)

This book contains information obtained from authentic and highly regarded sources. Reasonable efforts have been made to publish reliable data and information, but the author and publisher cannot assume responsibility for the validity of all materials or the consequences of their use. The authors and publishers have attempted to trace the copyright holders of all material reproduced in this publication and apologize to copyright holders if permission to publish in this form has not been obtained. If any copyright material has not been acknowledged please write and let us know so we may rectify in any future reprint.

Library of Congress Cataloging-in-Publication Data

Names: Qiu, Meikang, author. | Gai, Keke, author.
Title: Mobile cloud computing : models, implementation, and security /
Meikang Qiu, Keke Gai.
Description: Boca Raton : CRC Press, 2017.
Identifiers: LCCN 2017002243 | ISBN 9781498796033 (hardback : alk. paper)
Subjects: LCSH: Cloud computing. | Mobile computing.
Classification: LCC QA76.585 .Q58 2017 | DDC 004.67/82--dc23
LC record available at https://lccn.loc.gov/2017002243

Visit the Taylor & Francis Web site at
http://www.taylorandfrancis.com

and the CRC Press Web site at
http://www.crcpress.com

Dedications

We are enormously grateful to numerous individuals for their assistance in creating this book. First, we would like acknowledge those who have provided insights or feedback to this creation and the immeasurable help and support from the editors and anonymous reviewers. We also sincerely appreciate the support given by the Department of Computer Science at Pace University.

Dr. Qiu would like to thank his wife Diqiu Cao, son David Qiu, daughter Mary Qiu, father Shiqing Qiu, mother Longzhi Yuan, brother Meisheng Qiu, sister Meitang Qiu, and many other relatives for their continuous love, support, trust, and encouragement throughout his life. Without them, none of this would have happened.

Mr. Gai dedicates this work to his parents, father Jinchun Gai and mother Tianmei Li, who have brought him up and sacrificed so much. Dr. Gai could never have done this without his parents' love, support and constant encouragement. A sincere appreciation to all Keke's family members for their continuous love.

Contents

Part I **Basic Concepts and Mechanisms of Mobile Cloud Computing**

Chapter   1 ▪ Overview of Cloud Computing   3

Part III **Security Issues and Solutions in Mobile Cloud Systems**

Chapter   8 ▪ Security and Privacy Issues and Threats in MCC 151

List of Figures

List of Tables

Preface

This book focuses on introducing three vital aspects of mobile cloud computing, which are models, implementations, and security. We aim to assist graduate level students to learn the knowledge of mobile cloud computing for adopting the next generation technology of mobility, wireless networks, and application. The design of this book follows instructional principle in order to ensure the contents are reliable, teachable, and adoptable. Students can easily cognize concepts, models, and relevant applications throughout their study of this book. Moreover, this book is also for those students who seek for advanced algorithms applied in mobile cloud computing. A few novel algorithms in mobile cloud computing are covered in this book. The knowledge scope of this book can aid learners to obtain a panoramic and deep insight in mobile cloud computing.

Three main aspects of mobile apps are covered by this book, which represent three major dimensions in the current mobile app development domain. They are:

1. **Basic Concepts and Mechanisms of Mobile Cloud Computing.** This includes Chapters 1–4. We provide an overview of cloud computing by introducing concepts, service models, and service deployments in Chapter 1. Chapter 2 further introduces the specific cloud service models. Next, Chapter 3 presents basic mechanisms and principles of mobile computing, as well as virtualization techniques. Moreover, Chapter 4 introduces mobile cloud computing architecture design, key techniques, and main challenges.

2. **Optimization of Data Processing and Storage in Mobile Clouds.** Chapter 5–7 concentrate on the optimizations of mobile cloud computing. First, Chapter 5 introduces a few basic optimizations of the mobile cloud. Two aspects are addressed, which include performance and green clouds. Second, Chapter 6 covers the crucial optimization algorithm in mobile cloud computing,

which is preemptable algorithm. Finally, Chapter 7 addresses big data and service computing in mobile cloud computing.

3. **Security Issues and Solutions in Mobile Cloud Systems.** Chapter 8 and 9 concentrate on the security issues in mobile cloud computing. Among these two chapters, Chapter 8 provides a brief introduction about security and privacy issues and threats in mobile cloud computing. Chapter 9 further discusses privacy protection techniques in mobile systems.

4. **Integrating Service-Oriented Architecture with Mobile Cloud Computing.** Chapter 10 introduces the method of integrating Service-Oriented Architecture with mobile cloud systems. A few Web services specifications related to the implementations of mobile cloud computing are discussed in this chapter.

In summary, this book is written for those students/scholars who intend to explore the mobile cloud computing domain. All crucial aspects in mobile clouds are covered by this book. Moreover, this book not only presents critical concepts in mobile cloud systems but also drives students to advanced level studies. Some open discussion questions can facilitate students to do deeper research. Therefore, this is a handbook of mobile cloud computing that is suitable for a broad range of audiences.

About the Authors

Meikang Qiu received BE and ME degrees from Shanghai Jiao Tong University and a Ph.D. degree in Computer Science from the University of Texas at Dallas. Currently, he is an adjunct professor at Columbia University and Associate Professor of Computer Science at Pace University. He is an IEEE senior member and ACM senior member. He is the chair of the IEEE Smart Computing Technical Committee. His research interests include cyber security, cloud computing, big data storage, hybrid memory, heterogeneous systems, embedded systems, operating systems, opti-

mization, intelligent systems, and sensor networks. A lot of novel results have been produced and most of them have already been reported to the research community through high-quality journal and conference papers. He has published 12 books, 360 peer-reviewed journal and conference papers (including 160+ journal articles, 200+ conference papers, 60+ IEEE/ACM Transactions papers), and registered 3 patents. He has won the *ACM Transactions on Design Automation of Electrical Systems* (TODAES) 2011 Best Paper Award. His paper about cloud computing was published in the *JPDC* (*Journal of Parallel and Distributed Computing*, Elsevier) and ranked #1 in Top Hottest 25 Papers of *JPDC* 2012. He has won 8 other conference Best Paper Awards in recent years. Currently he is an associate editor of more than 10 international journals, including *IEEE Transactions on Computers* and *IEEE Transactions on Cloud Computing*. He is the general chair/program chair of a dozens of IEEE/ACM international conferences, such as IEEE HPCC, IEEE CSCloud, and IEEE BigDataSecurity. He has given more than 100 talks all over the world, including Oxford, Princeton, Stanford, and New York University. He won the Navy Summer Faculty Award in 2012 and the Air Force Summer Faculty Award in

2009. His research is supported by the US government: NSF, Air Force, Navy, and companies such as GE, Nokia, TCL, and Cavium.

Keke Gai holds degrees from the Nanjing University of Science and Technology (BEng), the University of British Columbia (MET) and Lawrence Technological University (MBA and MS). He is currently pursuing his PhD at the Department of Computer Science at Pace University, New York. Keke Gai has published more than 70 peer-reviewed journal or conference papers, more than 20 journal papers (including ACM/IEEE Transactions), and more than 40 conference papers. He has been granted three IEEE Best Paper Awards (IEEE SSC '16, IEEE CSCloud'15, IEEE BigDataSecurity'15) and two IEEE Best Student Paper Awards (IEEE HPCC '16, IEEE SmartCloud '16) by IEEE conferences in recent years. His paper about cloud computing has been granted the *Most Downloaded Article* of the *Journal of Network and Computer Applications* (JNCA) in 2016. He is involved in a number of professional/academic associations, including ACM and IEEE. Currently, he is serving as the Secretary/Treasurer of the IEEE STC (*Special Technical Community*) in Smart Computing at the IEEE Computer Society. He has worked for a few Fortune 500 enterprises, including SINOPEC and GE Capital. His research interests include mobile cloud computing, cyber security, combinatorial optimization, business process modeling, enterprise architecture, and Internet computing. He also served as Finance Chair/Operation Chair/Publicity Chair/Web Chair for serveral academic events, such as IEEE Smart-Com '16, NSS '15, and IEEE HPCC/ICESS/CSS '15.

Introduction

ABOUT THIS BOOK

Currently, developments in mobile technologies and wireless networks enable cloud-based services to be delivered by mobile devices, which is turning into a mainstream approach for deploying cloud services. However, there is no comprehensive textbook yet that clearly explains the mobile cloud computing model, its implementation, and security considerations, and is targeted for higher education.

The main motivator for writing a textbook concentrating on mobile cloud computing is to instruct graduate students majoring in Computer Science on the fundamental concepts, theoretical models, current implementations, security issues, and relevant techniques. These topics need to be addressed urgently at institutions of higher learning that aim to educate students to be successful as mobile computing professionals in the contemporary marketplace.

The audience for this textbook includes college lecturers at graduate schools who aim to teach masters-level students the essential theoretical knowledge concerning mobile cloud computing, supported by a series of practical case studies. The case studies will be designed to aid learners to understand the body of knowledge in a gradual and incremental way by means of a practical approach.

WHO SHOULD READ THIS BOOK

The audience for this textbook includes college lecturers at graduate schools who aspire to teach graduate students mobile cloud computing theories, using a series of practical case studies. The case studies will be designed for a learner-friendly delivery of the course material that utilizes practical applications of theoretical concepts. The potential market is large since most universities aim to include a graduate-level course focusing on mobile cloud computing in the curriculum; however, there is at the time of writing no comprehensive textbook available.

This is an opportune moment to publish a textbook with the proposed intent and concentration.

EACH CHAPTER'S CONTENT OVERVIEW

Chapter 1: Overview of Cloud Computing

Introducing key concepts of cloud computing, service deployments, models, and architecture, presenting current adoptions of cloud computing and performance, underlying principles and methodologies, and reviewing advantages and disadvantages of cloud computing implementations.

Chapter 2: Specific Cloud Service Models

Introducing more specific service models in order to assist students in understanding the implementations of cloud computing in different scenarios. The involved specific cloud service models in this chapter include *Desktop-as-a-Service, Storage-as-a-Service, Database-as-a-Service, Backend-as-a-Service, Information-as-a-Service, Integration-as-a-Service, Security-as-a-Service,* and *Management/Governance-as-a-Service.*

Chapter 3: Basic Mechanisms and Principles of Mobile Cloud Computing and Virtualization

Introducing development of mobile cloud computing and the key techniques, describing the nature of mobile cloud computing, identifying advantages and disadvantages, describing basic considerations of deploying mobile cloud computing. Introducing the main mobile techniques implemented in practice and the key features, defining wireless networking characteristics and ways of development, describing the main operating systems, and reviewing the different generations of mobile devices. Introducing mobile Internet and its characteristics, including Web services, wireless networks, and key techniques; describing the evolution of mobile Internet, detailing wireless access and prevailing standards.

Chapter 4. Mobile Cloud Computing Architecture Design, Key Techniques, and challenges

Understanding the application architecture for leveraging mobile cloud computing, describing information flows and business processes, introducing design methodology of applications, and describing the migration to mobile clouds. Describing virtualization implementations and applications in mobile cloud computing, identifying the importance of virtualization and listing the pros and cons of virtualization.

Chapter 5: Basic Optimizations: Cloud Computing Performance and Green Clouds.

Performance of cloud computing and green clouds are two crucial issues in implementing cloud-based solutions. There are many different parameters that can be considered when cloud designers design a cloud system. Students need to understand the crucial aspects of cloud computing performance in the development of cloud systems after reading this chapter.

Chapter 6. Preemptable Algorithm Execution in Mobile Cloud Systems

Online optimizations using scheduling of preemptable tasks in cloud systems is an important approach for increasing the efficiency of cloud systems. The goal of this optimization algorithm is to improve the performance of the clustered distributed cloud remote servers. The total cost of using mobile clouds will be reduced by executing this algorithm in a pool of cloud computing resources.

Chapter 7. Big Data and Service Computing in Cloud Computing

Big data and service computing in cloud computing is an important aspect of service deliveries. This chapter focuses on these two concentrations: the first focus is big data and service computing in cloud computing; the other focus is phase-reconfigurable shuffle optimizations for MapReduce in cloud computing. Reading this chapter can assist students to not only have a basic picture about big data implementations in cloud computing, but also learn the advanced algorithms of MapReduce in the cloud context.

Chapter 8. Security and Privacy Issues and Threats in Mobile Cloud Computing

Security and privacy are significant aspects for mobile cloud users, developers, and service vendors, since any privacy leakage may result in serious unexpected consequences. Currently, users' sensitive information is facing various threats, some of which are caused by the implementations of Web or mobility technologies.

Many security problems in mobile cloud computing are generally associated with privacy issues. In this chapter, we summarize and review major security and privacy problems and introduce the main taxonomy of threats in mobile cloud computing.

Chapter 9. Privacy Protection Techniques in Mobile Cloud Computing

Techniques of protecting security and privacy are critical aspects in securing services and operations in mobile cloud computing. In Chapter 8, we talked about basic concepts in security and privacy issues and threat models. Students should have an overview of security and privacy in mobile cloud computing after reading Chapter 8. This chapter focuses on a number of crucial security dimensions to assist students in understanding the specific security problems. A few security solutions will be introduced in this chapter, too.

Chapter 10. Web Services in Cloud Computing

Web services is fundamental in delivering cloud computing services. The device-to-device services also empower the implementations of *Service-Oriented Architecture* (SOA). In this chapter, we will discuss the relationships among cloud computing, SOA, and Web services, as well as explain the significance of integrating SOA with mobile cloud computing. Several Web services specifications will be introduced in this chapter as well.

I

Basic Concepts and Mechanisms of Mobile Cloud Computing

Overview of Cloud Computing

CONTENTS

Cloud COMPUTING has become one of the buzziest words in computing industry and a way of life for people. As an emerging technology, mobile cloud computing has provided mobile users with a variety of new service approaches using various manners, from integrating multiple techniques to exploring novel technologies. It is a technology converging a few technologies from multiple fields, including mobile technologies, mobile networks, and cloud computing. The principle of applying mobile cloud computing is to gain the benefits of cloud computing within a wireless implementation environment. Being aware of basic concepts of cloud computing is fundamental in understanding mobile cloud computing. In this chapter, we are reaching the core aspect of mobile cloud computing technology, which is cloud computing. We want students to acquire a brief picture about cloud computing after reading this chapter. It is very important for students to correctly understand the operating principles of cloud computing before students explore the domain of mobility. The methodologies of adopting cloud computing will provide students with fundamental knowledge regarding the cloud's adoptions and implementations, which covers the following aspects:

1. Key concepts of cloud computing

2. Service deployments, models, and architecture

3. Presenting current adoptions of cloud computing and performance

4. Underlying principles and methodologies

5. Reviewing advantages and disadvantages of cloud computing implementations

After reading this chapter, students should be able to answer the following questions:

1. What are the key concepts of cloud computing?

2. What are cloud computing service deployments?

3. What are cloud computing service models?

4. What are the main cloud computing architectures?

5. What are key techniques behind the clouds?

6. How does current cloud computing perform?

7. What are the main principles and methodologies of cloud computing?

8. What are the main advantages and disadvantages of adopting cloud computing technologies?

1.1 INTRODUCTION

This chapter will give students an overview of cloud computing with an understanding of basic concepts, principles, and implementations. Some concepts will be used throughout this book, such as service deployments and service models. Following the instructions of this chapter, students will have a clear cognition of the architecture in which cloud-based services are adopted. Being aware of techniques behind the clouds can assist students to further understand the operations or techniques behind mobile cloud computing, which will be introduced in the following chapters.

1.2 CONCEPT OF CLOUD COMPUTING

The concept of *Cloud Computing* has been discussed by researchers over the years and many scholars have asserted a variety of concepts of cloud computing [3, 4, 5, 6]. Many considered cloud computing as a new concept rather than a new technology. Based on the prior attempts, it is still difficult to formulate a general recognized concept because of its wide usages and a broad scope of computing technologies and relevant resources.

In this book, we define *Cloud Computing* as a type of Internet-based computing that provides users with multiple scalable on-demand services, data, or products through sharing or accessing various computing resources. Those resources can from private systems or from third-party data centers, and can be accessed locally or remotely from city-wide to world-wide. Using cloud means users can elastically obtain hardware, software, or other computing resources on demand. As a scalable and flexible service solution, cloud computing combines a new paradigm with existing technologies. This definition will guide the rest part of the book to ensure all audiences have the same understanding of cloud computing when exploring mobile cloud computing.

The term Cloud is a metaphor for describing the servers providing services via the Internet that can be hosted or maintained by any third party [3]. The server on the cloud is the core of cloud computing and currently it can support most types of computing resources as services. Compared with the classic rented servers, cloud computing provides more flexible and feasible solutions, which depend on users' demands and used computing resources [7, 8]. Cloud service providers maintain the cloud-based servers and handle the technical problems. Users purchase computing services from the cloud service providers so that users can acquire more benefits from utilizing computing resources and technologies without the investment in technology development. This principle is related to most characteristics of cloud computing.

1.3 CHARACTERISTICS OF CLOUD COMPUTING

Cloud computing also exhibits some characteristics that are generally accepted by scholars and practitioners. A few key characteristics include masked complexity, self-service demand, advanced flexibility, broad network access, resource pooling, risk migration, and measured service. A cloud-based solution may have some or all of these features.

1. **Masked Complexity** This feature is one of the most important characteristics of adopting cloud computing in practice. It refers to the cloud users' ability to leverage cloud-based solutions to quickly acquire value via sophisticated services or products while the complexities are masked behind the user interface by the support of cloud service providers.

2. **Self-Service Demand** This characteristic means cloud users can determine the scope of services by using a user interface provided by cloud service providers, such as an online control panel. It enables cloud users to design their service requirement in a flexible and economic way and scale services up and down depending on the in-time demands.

3. **Broad Network Access** This property means the cloud-based services are available on multiple platforms simultaneously, such as a desktop and various mobile devices.

4. **Advanced Flexibility** This characteristic is also known as *Rapid Elasticity* [9], which refers to a higher-level capability of scalable services. Distinguishing from *Self-Service Demand*, this essential aspect of cloud computing means that cloud providers are responsible for the provision of scalable computing resources.

5. **Multi-Tenancy Principle** This feature applies to those cloud providers who serve multiple customers using a multi-tenancy principle. In this principle, a tenant refers to a group of customers who share the same view of the applications. Leveraging

this principle can usually assist cloud providers to increase utilization and efficiencies of computing resources by a number of approaches, such as centralizing physical locations of infrastructure and optimizing workload level.

6. **Risk Mitigation** The term *Risk* covers two security issues while adopting cloud computing. One is to migrate security concerns to cloud service providers who will protect computing resources from attacks and other threats. The other side is to avoid the risk of technical development by acquiring computing resources from cloud providers.

7. **Measured Service** Most costs of cloud-based services are calculable so that customers can determine which cloud services they want to purchase and how long they want to use the services. The usage of computing resources should be trackable, controllable, and monitorable by both service providers and consumers.

In a perspective of practice, a cloud-based solution may have some other characteristics, such as cost reductions, accessibility by leveraging *Application Programming Interfaces* (APIs), easy maintenance for users, and measured performances, productivity, and reliability. These characteristics may exist in some scenarios of adopting cloud computing, which is influenced by the circumstances or other objective elements, such as customers' requirements, market needs, technical burdens, service provider policies, and legal issues.

Cloud computing is important for current enterprises because there are some facets distinguishing cloud computing from other computing models [10], such as on-demand self-services, high adaptability, and flexibility of leveraging technology-based solutions without the use of hosting servers. These advantages are achieved through providing various types of cloud services. Advantages and disadvantages may be varied due to the distinctions of service models and service deployments. Understanding both benefits and drawbacks for each cloud service model and service deployment is significant for adopting cloud computing technology in practice, which we will discuss more in this chapter.

1.4 BASIC CLOUD COMPUTING SERVICE MODELS

A *Service Model* in cloud computing refers to an agile approach of delivering specific services that can properly meet the customers' de-

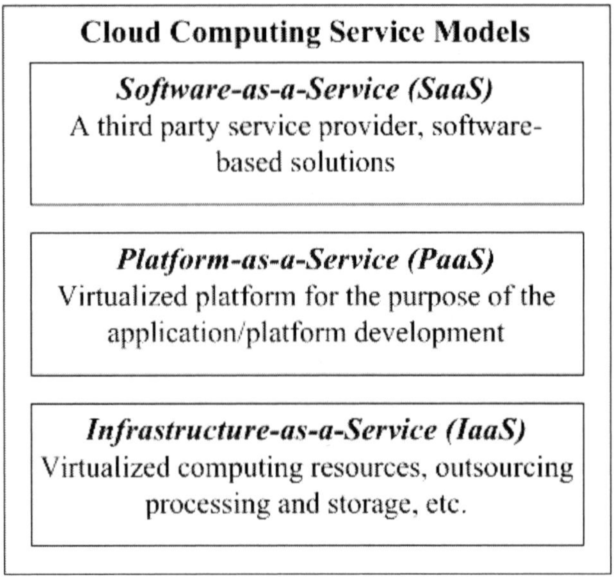

```
┌─────────────────────────────────────────────────┐
│          Cloud Computing Service Models           │
│  ┌───────────────────────────────────────────┐   │
│  │        Software-as-a-Service (SaaS)         │   │
│  │     A third party service provider,         │   │
│  │        software-based solutions             │   │
│  └───────────────────────────────────────────┘   │
│  ┌───────────────────────────────────────────┐   │
│  │        Platform-as-a-Service (PaaS)         │   │
│  │   Virtualized platform for the purpose of   │   │
│  │   the application/platform development       │   │
│  └───────────────────────────────────────────┘   │
│  ┌───────────────────────────────────────────┐   │
│  │      Infrastructure-as-a-Service (IaaS)     │   │
│  │  Virtualized computing resources,           │   │
│  │  outsourcing processing and storage, etc.   │   │
│  └───────────────────────────────────────────┘   │
└─────────────────────────────────────────────────┘
```

Figure 1.1   Cloud computing service model layers.

mands. There are many different service models being implemented in the current industry due to various market needs. The method of categorizing service models is usually determined by the classifications of services offered by the cloud providers, which can also be considered a layer concept. This system groups cloud computing service models according to the objects that are providing services for customers. We mainly introduce three fundamental service models of cloud computing in this section, including *Infrastructure-as-a-Service* (IaaS), *Platform-as-a-Service* (PaaS), and *Software-as-a-Service* (SaaS). Other cloud service models, also known as *Specific Cloud Service Models*, will be introduced in the following section. Figure 1.1 illustrates a basic layer structure of cloud computing service models.

1.4.1   Infrastructure-as-a-Service

Infrastructure-as-a-Service (IaaS) is a service model that enables end-users to acquire virtualized computing resources, such as hard drives, processors, and memory cards, from cloud providers [7, 11]. As shown in Figure 1.1, IaaS is the first layer of the layer structure. It implies that IaaS is the foundation of cloud computing. Using this service model, customers are able to focus on operating systems, applications, data, and other operations related to their business since cloud service providers are responsible for all the work behind the scenes, such as maintaining servers, networks, processing, and storage.

An example of implementing IaaS is a situation where an organization needs to acquire some data storage space in a dynamic manner depending on market needs. Leveraging the IaaS model can allow the organization to scale the size of the purchased data storage service up and down in-time demands.

Figure 1.2 displays the high level architecture of adopting IaaS that represents the scenario proposed in the example. As illustrated in Figure 1.2, *Cloud Users* can determine the service scopes and scales by their policies and demands. Service requests and service responses are delivered by Internet connectivity. The service is represented by the *Desktop Virtualization* that allows the client to access to the service by an isolated virtualized desktop environment.

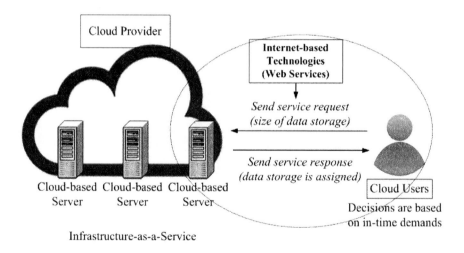

Figure 1.2 An example of high-level *Infrastructure-as-a-Service* (IaaS) architecture.

The main advantage of IaaS is allowing enterprises to access to the costly computing resources within a required scope. The computing resources could be database, data storage, processors, memories, routers, networks, and other virtualized physical infrastructure. Cloud service providers can supply same-level services to multiple clients simultaneously in order to maximize the capability of the infrastructure, such as hard drive. Meanwhile, clients can purchase services from diverse cloud service providers in order to meet their business requirements.

Moreover, IaaS is an *Internet Protocol* (IP)-based model that delivers services regardless of the geographic location of the infrastructure [12]. Customers are not aware of the exact locations of the physical infrastructure that is providing customers with the services. Without worrying about the technical complexity and geographic restrictions, cloud users are able to pay attention to other aspects that are valuable to their businesses, such as business process improvement and strategy-making. This benefit is also available in the other two fundamental service models, PaaS and SaaS.

1.4.2 Platform-as-a-Service

The second basic cloud computing service model is *Platform-as-a-Service* (PaaS), which is a service model that allows web developers to utilize a complete virtualized platform for the purpose of application or platform development. This service model consists of a number of subsystems and interfaces within a common structure, involving a stream of relevant product development procedures. The *Platform* is a group of subsystems that meets all the requirements for the application development and hosting operations, including interface development, database development, data storage, system testing, and other processes related to application development. Subscribers can easily build enterprise-level applications in a customized manner following the enterprises' strategies and policies, such as a Website or a mobile app.

An important feature of PaaS is to provide external developers with an open Web-based development platform [13, 9]. This feature enables those small or medium-size enterprises that have limited computing resources to explore their markets by leveraging advanced computing techniques according to their needs.

For example, leveraging PaaS-based solutions can assist a global service organization to add value to the existing services and create new services, which has been examined by the prior research. This example assumes an *Enterprise A* is an American global service organization that facilitates study abroad offerings for foreign students, such as Chinese students. One solution for Enterprise A to expand the international markets is to develop and operate a portal site with multiple online services. In this case, leveraging PaaS is an efficient approach for

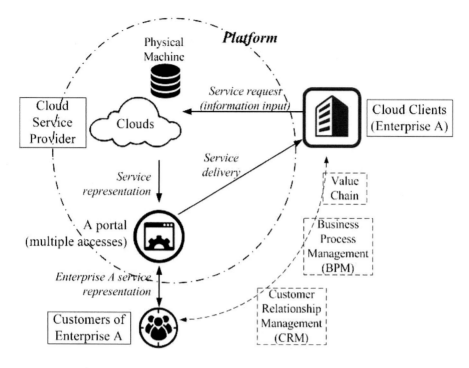

Figure 1.3 An example of *Platform-as-a-Service*.

Enterprise A to customize their online services by developing a portal, depending on their business strategies and goals.

Figure 1.3 represents an example of adopting the PaaS implemented in the scenario provided in the example. As illustrated in Figure 1.3, Enterprise A leverages a PaaS-based solution to create a portal with multiple accesses to reach the target customers. The services are represented by the portal that is developed and customized according to the business demands of Enterprise A, such as improving value chain, *Customer Relationship Management* (CRM), and *Business Process Management* (BPM).

An efficient PaaS solution should have some or all of these aspects as the core values of using PaaS solutions. The core values are listed as follows:

1. **Design** The platform should allow users to customize the *User Interface* (UI) design.

2. **Development** Users should be able to develop the application

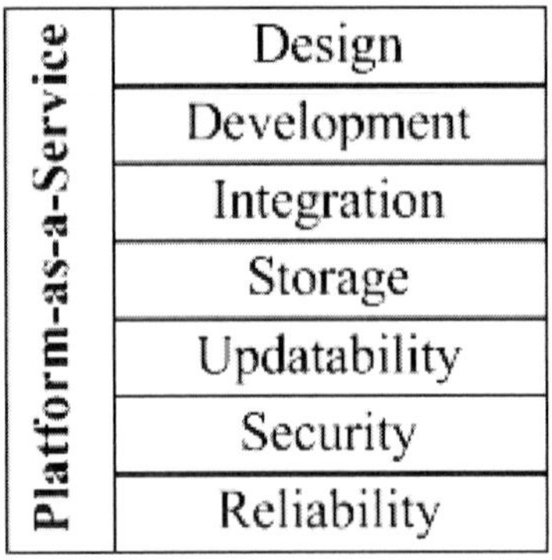

	Design
	Development
Platform-as-a-Service	Integration
	Storage
	Updatability
	Security
	Reliability

Figure 1.4 Core value of *Platform-as-a-Service*.

with no restriction of programming skills and software development knowledge.

3. **Integration** The platform should support multiple applications simultaneously. Meanwhile, multiple services can also be integrated on the platform, such as storage, testing, and applications.

4. **Storage** Persistent data storage supporting applications should be available for users, which includes both on-demand database and on-demand file storage. It is commonly associated with the commercial contract.

5. **Updatability** The platform needs to support on-demand updatability according to users' requirements.

6. **Security** Users' information must be protected from threats that can be either external attacks or internal improper operations.

7. **Reliability** Cloud service providers need to ensure the platform service can last long and be continuously well maintained.

These aspects can also be considered benefits of PaaS in the perspective of operations. Although there are a number of advantages, there are two common drawbacks of implementing PaaS-based solutions. First, most PaaS solutions are usually available only on subscription, which may cause some trouble in service provider selections and platform tests. Second, it will be difficult for users to switch from one platform to another platform since most *Platforms* cannot be integrated or communicated. Cloud clients need to carefully consider these two issues before they officially migrate to Platform-as-a-Service.

1.4.3 Software-as-a-Service

Software-as-a-Service (SaaS) is also known as *Application-as-a-Service*, which refers to leveraging cloud-based software solutions to provide users with web-based software, platform, and infrastructure services [14]. Similar to IaaS and PaaS, the hardware or software can be hosted by a third party so that users may determine the content of the services depending on their needs. Generally, users access SaaS via a Web browser by utilizing a thin-client computer terminal. The most prominent feature of SaaS is that clients can access enterprise-class applications without purchasing and installing enterprise software [15]. As exhibited in Figure 1.1, SaaS is on the top of the layer structures, which implies that SaaS is utilized on an operational level.

SaaS can satisfy a wide range of demands due to the diversity of software. The software can be either an enterprise-level system or a personal-level application. For example, SaaS can deliver large services, such as *Enterprise Resource Planning* (ERP), or a smaller sized application for individual users, such as an office word processing tool.

With a broad service scope, SaaS is a service model that usually offers low-cost features to customers. Leveraging the multi-tenant architecture can lower the cost since one configuration can serve all tenants. Additionally, instead of buying an application license, SaaS users purchase services by a subscription fee that is typically based on usage. The subscription fee can be counted monthly, quarterly, yearly, or on other flexible plans. It can further make the cost lower since an on-demand purchase is usually cheaper than a full payment in advance. This model has a great influence on large applications.

Moreover, the concept of SaaS is a complicated architecture offering enterprises advanced flexibility and extensibility for the purpose of handling changes in usage rates and APIs, rather than simply uploading software on the Web [16]. Behind the Web browser, SaaS is supported by a variety of mechanisms. Two common schemas are the multi-tenant architecture and virtualization.

In essence, besides the characteristics of cloud computing, an SaaS-based solution may have other characteristics because of the direct connections with consumers. The following list describes characteristics of SaaS applications. An SaaS application may have some or all of these features.

1. **Update** Only service providers can update the applications.

2. **Configuration** There is usually only one configuration for the application.

3. **Integration Protocol** SaaS applications offer integration protocols that are mainly based on *Hypertext Transfer Protocol* (HTTP), *Simple Object Access Protocol* (SOAP), *Representational State Transfer* (REST), and *JavaScript Object Notation* (JSON).

4. **Access** SaaS vendors can access the database by authorizations and acquire part of data for business purposes.

1.5 CLOUD SERVICES DEPLOYMENTS

Cloud Service Deployment is a term describing service implementation methods for utilizing cloud computing-based solutions. Current cloud services exist in four types, namely public cloud, private cloud, community cloud, and hybrid cloud. These four types of services are used to describe the cloud infrastructure's physical arrangement and the manner of receiving cloud services by the end users.

1.5.1 Public Cloud Computing

A public cloud is hosted or operated by a service provider who sells or offers a multitude of services to the public. This service deployment usually provides end users with a scalable on-demand service. Examples of current public cloud providers include Amazon, Google, Rackspace, and Salesforce. Leveraging the public cloud has a number of limitations for end users but it has an advantage in flexibility and on-demand self-service. Security and privacy are two of the most common concerns for users, and are relevant to data service outsourcing, computation outsourcing, access control, and trustworthy service metering. Other limitations of leveraging the public cloud for business applications include absence of service-level agreements, risks of interoperability, instability of performance, limits of latency and network, and non-scalability of storage.

1.5.2 Private Cloud Computing

A private cloud is owned or leased by an individual organization and is only operated or served for the private use of the individual organization. Enterprises select this cloud deployment when they want to deploy the cloud infrastructure inside the company'ĂŹs firewall and operate the systems within an in-house cloud.

Moreover, this deployment model assists enterprises in reducing the total cost of servers and power by centralizing their in-house systems. Compared with the public cloud, the aim of a private cloud is to provide internal users with a flexible and agile private infrastructure rather than selling cloud services to the public.

1.5.3 Community Cloud Computing

The community cloud is a deployment model that enables several organizations to host cloud infrastructures and sell or offer cloud services to a specific group of organizations. Therefore, the cloud-based computing resources are shared within a community with similar interests.

This model is similar to the private cloud in management but targets a specific set of customers. Two main concerns of using the community cloud are cost and data safety. The cost of the community cloud is usually higher than a public cloud and bandwidth and data storage are shared with all community members.

1.5.4 Hybrid Cloud Computing

The hybrid cloud is a deployment model that merges two or more cloud service deployments, such as the private, public, and community clouds. A hybrid cloud consists of at least one private cloud and at least one public cloud. There are two common ways of providing a private and a public cloud in a hybrid cloud.

The first method is that an end user hosting a private cloud forms a partnership with a public cloud provider. The other method is that an end user forms a partnership with a public cloud provider who provides the end user with a private cloud service. The main purpose of operating both private and public clouds is to have control over the critical data and infrastructure within a private computing environment and outsource other non-critical services from the public cloud providers [10]. The main risk of a hybrid cloud is inherent in the public cloud. Fully understanding these four cloud service deployments can assist enterprises that are willing to adapt cloud computing to determine an appropriate deployment that fits their business. Two main concerns in deciding on these deployments are cost level and security issues [17].

Private cloud computing provides a higher-level secure computing environment, although it usually costs more than other deployments. On the contrary, security issues become a main concern when enterprises use public cloud computing. A community cloud is positioned as a mixture of private cloud and public cloud processes which may aid a community to obtain a safe cloud service with a lower cost. Hybrid clouds enable enterprises to store their regular data and sensitive data in public and private cloud servers, respectively.

1.6 SUMMARY

We went through a series of important concepts and aspects in cloud computing, from basic definitions to common cloud computing service models. Students should understand the features of cloud computing from reading this chapter, which were masked complexity, self-service demand, broad network access, advanced flexibility, multi-tenancy principle, risk mitigation, and measured service. We also talked about three basic cloud computing service models, which were Software-as-a-Service, Platform-as-a-Service, and Infrastructure-as-a-Service. Moreover, there were four fundamental cloud service deployments: public, private, community, and hybrid cloud computing.

1.7 EXERCISES

Students should be able to answer the following questions after reading Chapter 1. The goals of these practices are to assist students in fully understanding the concept of cloud computing and its service models and service deployments.

1. What is the concept of cloud computing?

2. How do you understand the term "cloud" in cloud computing?

3. What are common characteristics of cloud computing?

4. Why is *Masked Complexity* an important characteristic for implementing cloud solutions?

5. What are the benefits of *Demand Self-Service*?

6. What does *Broad Network Access* mean?

7. What is the principle for cloud providers to support multiple customers who have the same views on the applications?

8. What is your understanding of the term *Risks* when adopting cloud computing solutions?

9. What is a *Service Model*?

10. What are three basic cloud service models? What is a *Layered Approach*?

11. What are main advantages of adopting IaaS?

12. What does term *Platform* mean in PaaS? What are core values of leveraging PaaS solutions?

13. What are features of SaaS besides the characteristics of cloud computing?

1.8 GLOSSARY

Cloud Computing leverage Web-based technologies to provide users with multiple scalable on-demand services or products for sharing or offering various computing resources.

Cloud Service Deployment is a term describing service implementation methods for utilizing cloud computing-based solutions.

Cloud Service Model is an approach of delivering specific services that can properly meet customers' demands.

Infrastructure-as-a-Service (IaaS) is a service model that enables end users to acquire virtualized computing resources, such as hard drives, processors, and memory cards, from cloud providers.

Platform-as-a-Service (PaaS) is a service model that allows web developers to utilize a virtualized platform for the purpose of application or platform development.

Software-as-a-Service (SaaS) leverages cloud-based software solutions to provide users with web-based software, platform; and infrastructure services; also known as Application-as-a-Service.

Specific Cloud Service Models

CONTENTS

Specific CLOUD SERVICE MODELS have driven flexible service offerings in cloud computing. Excluding Infrastructure-as-a-Service, Platform-as-a-Service, and Software-as-a-Service, there are other specific cloud service models in practice that have been developed for some specific business targets. These specific cloud service models are related to or overlap with each other for various purposes or matching distinct demands.

1. Why do we need *Backend-as-a-Service*?

2. What are differences among *Backend-as-a-Service*, *Storage-as-a-Service* and *Database-as-a-Service*?

3. How is *Security-as-a-Service* implemented?

4. What is *Management/Governance-as-a-Service*?

Some specific cloud service models are *Desktop-as-a-Service*, *Storage-as-a-Service*, *Database-as-a-Service*, *Backend-as-a-Service*, *Information-as-a-Service*, *Process-as-a-Service*, *Integration-as-a-Service*, *Security-as-a-Service*, *Management/ Governance-as-a-Service* (MaaS and GaaS), *Information -Technology -Management -as-a-Service* (IT-MaaS), and *Testing-as-a-Service* (TaaS).

Current enterprises that intend to use cloud-based services can determine a proper service model depending on the enterpriseĂŹ strategies, demands, business models, financial conditions, and security concerns [18, 19]. In order to successfully address an enterprise's information technology strategy, that enterprise must understand which cloud service deployment matches its demands.

2.1 DESKTOP-AS-A-SERVICE

Desktop-as-a-Service is a cloud computing solution providing cloud customers with an approach of using distributed computing resources by virtualizing remote device desktops and functions. Those cloud customers who have limited computing resources but require distributed working environments can leverage Desktop-as-a-Service offerings to increase their computing facility's utilizations. The service is implemented as an application-based solution that represents a virtual desktop to enable cloud customers to control or operate their devices remotely. The typical example is using a *Virtual Machine* (VM) in executing cloud computing service, which is one of the main techniques for implementing cloud computing. A detailed introduction to VM will be given in the succeeding chapters.

Figure 2.1 represents a framework of Desktop-as-a-Service. As exhibited in the figure, dotted lines with arrows refer to the operations of cloud users and work flow directions. Broken lines with arrows stand for service delivery processes. A Desktop-as-a-Service application is a *User Interface* (UI) in which cloud users send out service requests and

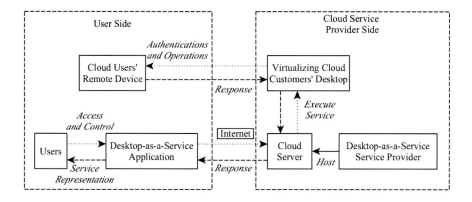

Figure 2.1 Framework of Desktop-as-a-Service.

receive the service representation. A typical service representation is an operating desktop interface that virtualizes cloud users' remote devices. Cloud service providers physically host the back-end of the virtual interface infrastructure. Information on the back-end can be found in Section 2.4.

2.2 STORAGE-AS-A-SERVICE

2.2.1 Main Concepts

Storage-as-a-Service is a cloud-based architectural solution that provides end users with data storage services, which is often considered a business model [20, 21, 22]. It is a primitive approach of adopting cloud-based services for those local applications that require a larger storage capacity than the local equipment physically has. The storage is available remotely provided by cloud providers who physically host the data storage facility. User applications can leverage the storage by purchasing the service according to their real-time demands.

Figure 2.2 represents a framework of deploying Storage-as-a-Service. Applications leverage both local disks and remote disks. The remote disk extends the role of the local disk and offers a larger storage volume. Users do not need to know the physical locations where the storage facility is deployed. The extended capacity can be used for multiple purposes, such as extra data storage, database backup, data synchronizations, data shares or exchanges, and data protection.

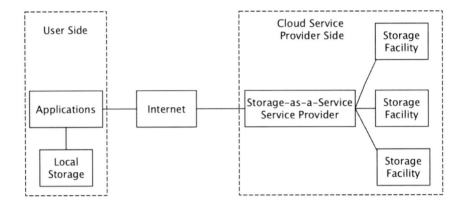

Figure 2.2 Framework of Storage-as-a-Service.

2.2.2 Benefits and Drawbacks

As a cloud-based service, Storage-as-a-Service requires Internet access. User applications connect with a remote disk drive, over the Internet, in which data are physically stored. Leveraging Internet-based connections enables users to have a flexible plan for disk usage. There are a number of crucial benefits of adopting Storage-as-a-Service.

1. *Flexibility and Lower Costs* Storage-as-a-Service enables users to have a flexible plan for using, managing, and upgrading their database usage. Enterprises or individual users can purchase the service depending on their requirements of the storage capacity in a dynamic manner. Cloud providers usually provide cloud customers with a scalable service that allows the customer to adjust the remote disk size based on their needs.

   This type of service can help cloud customers avoid over-maintenance of storage infrastructure. Cloud users can decrease the remote disk space when the demand is reduced, which is associated with saving costs.

2. *Usability and Accessibility* Users adopting Storage-as-a-Service do not need to maintain a great volume of storage infrastructure. Compared with local storage methods, cloud-based storage services provide users with better performance in stability and continuity. Cloud service providers take responsibilities for repairing all kinds of mechanical failure. Users are able to save money, energy, and time by avoiding costly technical consumption and machine crashes.

Moreover, technical support from cloud service providers can help cloud users achieve multiple methods for accessing the data in the clouds. For example, data in the cloud are accessible via either wired networks or mobile networks. Users do not necessarily carry a specific storage hardware when they need to access data. All data are available when users have access to the Internet and an authentication to the disk.

3. *Security, Protection, and Recovery* Generally, Storage-as-a-Service providers offer the same level security as the localized storage does. Sensitive and private information stored in the clouds can be protected in several ways, such as encryptions, secure protocols, user authentications, and traffic surveillance.

   Furthermore, cloud-based storage is considered a higher-level security storage approach in some situations. Many Storage-as-a-Service providers have backups for each consumer in case of local data loss or system crashes. The backup service is an extra layer for enhancing the data security level, compared with pure local storage.

4. *Other Storage-Based Services* Current Storage-as-a-Service also offers some additional services related to storage. Storage shares and synchronizations are two common plus services in Storage-as-a-Service. First, assigning multiple user authentications can enable storage shares. The authorized users can share access to a disk, a folder, or a file in the cloud synchronously. Next, a synchronization service means that data are automatically synced to the cloud and are available on multiple devices. The synchronization service can make the last edited or added file available on all devices by using the same or shared authentications.

Despite a few benefits, there are still some limitations when adopting Storage-as-a-Service.

1. Users must have an Internet connection when they need to acquire data from cloud storage. This may cause a few issues when losing access or suffering a weak connection.

2. Security issues can be another concern when leveraging sharing and syncing functions. Files shared by multiple users may face more risks because of various devices and heterogeneous networks. Threats may occur when user authentication fails.

3. The performance of networks may influence the service quality. Restrictions caused by slow networks can cause trouble for those applications that require higher performance.

4. Storage-as-a-Service is beneficial for those users who require distributed document sharing and syncs. The costs can be reduced by saving on the expenses for hiring technicians and maintaining storage space. However, this service method will not bring much difference if distributions are not needed.

2.3 DATABASE-AS-A-SERVICE

2.3.1 Main Concepts

Database-as-a-Service is a scalable service that provides remote database capacity as if the database is hosted locally. The remote database can supply multiple users with a service presentation showing the same database functions as the on-premises solution. This service model targets both professionals and those users who do not have a technical background or database development skills. Professionals who may be developers, database administrators, or system architects can easily obtain a full functional database in the cloud for testing, managing, designing, or extending services. On the other side, non-technical users can build their own on-demand databases without learning much knowledge about databases. Service providers usually supply cloud consumers with a flexible service option that is easily established, used, and configured. This feature is important for business because enterprises can deploy or upgrade databases for a fast response to market needs.

Figure 2.3 represents the framework of Database-as-a-Service. Local applications need to connect to the database in the cloud to upload and download data. Database-as-a-Service providers can leverage distributed databases to support multiple tenants.

2.3.2 Benefits and Drawbacks

Besides most benefits of cloud computing, leveraging Database-as-a-Service has a number of characteristic advantages in database governance.

1. *Optimized Database Administration* Enterprises or individual users can focus on the higher level of database administration instead of tedious and repetitive work. Outsourcing database

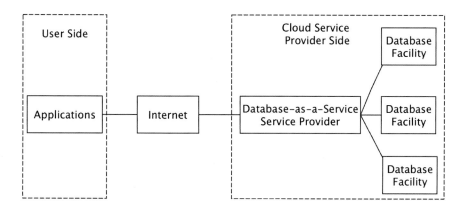

Figure 2.3 Framework of Database-as-a-Service.

administration and maintenance enables enterprises to pay more attention to optimizing database logics. Enterprises can pass an efficient database requirement to service providers to guide the target performance.

Database-as-a-Service provides an automatic database administration and surveillance. Migrating administration to the cloud can shorten the database establishment time as well.

2. *Effective Cost Planning* Currently, hosting a database is costly due to the high costs of technicians, maintenance fees, and database licenses. By acquiring the same functionality as the on-premises databases, cloud users can determine the scale of database usage on demand. The criterion of usage costs can be determined by multiple parameters, such as backup, recovery, table, timing, accessibility, computing resource, and other database functionalities.

Similar to Storage-as-a-Service, Database-as-a-Service also has a few disadvantages.

1. Database-as-a-Service often faces a security concern when high-level security is required. The database in the cloud is run within a virtualized container that can logically ensure the security of the database. However, users do not directly control the remote disks, so the implemented security technologies are ambiguous for users.

The private and sensitive information stored on cloud databases are protected under the methods offered by the service providers. In practice, users need to discern the security methods used by cloud providers when a full security requirement is needed. Based on the security technologies, users need to generate their own database security management strategies and methods.

2. Performance can also be a restriction to adopting Database-as-a-Service. A database running in the cloud is a service representation by virtualizing database functions. However, virtualization cannot achieve exactly the same capability and functionality as on-premises hardware in most situations. Users need to accept the outcomes of missing some database functions.

Box 2.3.2: Comparison between Database-as-a-Service and Storage-as-a-Service.

Database-as-a-Service differs from Storage-as-a-Service in a few aspects.

First, Database-as-a-Service offers a higher level of functionality, performing like a local database. Storage- as-a-Service only has service offerings related to storing data. Second, Database-as-a-Service and Storage-as-a-Service have different target markets that address different demands. Database-as-a-Service mainly focuses on migrating database administration and maintenance into the cloud so that the services are typically leveraged by IT technicians, such as a database administrator and networking risk manager. On the contrary, Storage-as-a-Service can be leveraged by all employees working at the enterprise or individual users who require additional disk space.

Comparisons between Database-as-a-Service and Storage-as-a-Service are given in Box 2.3.2.

2.4 BACKEND-AS-A-SERVICE

Backend-as-a-Service (BaaS), also known as *Mobile Backend-as-a-Service*, is a new service model that connects applications with required backend cloud computing resources and *Application Programming Interfaces* (APIs). The *Backend* in the concept refers to an application

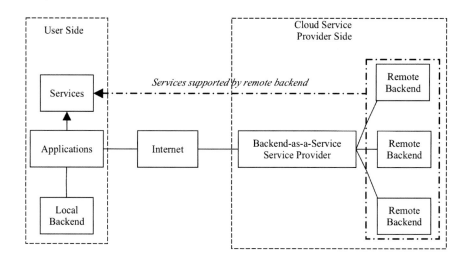

Figure 2.4 Framework of mobile Backend-as-a-Service.

or program hosted on the remote disk by which service representation is supported. BaaS plays a mediator role between front applications and backend supportive programs in the cloud. The principle of operating BaaS is that various applications may require the same or similar programs when developing web-based applications. Developers do not need to develop their own backend for some specific functions and save time by avoiding repetitive programming work. It can be considered a "plug-in" development model that helps developers efficiently build the required backend by using remote computing resources.

Currently, BaaS is widely used for efficient-response app development in mobile cloud computing. Leveraging the packed tools can assist cloud users to easily achieve a number of services that require backend, which the apps need to access. Cloud service providers offer cloud users a packed set of development tools and computing resources, often called *Custom Software Development Kits* (CSDK), which run on cloud-based servers.

Figure 2.4 represents a framework for mobile BaaS. Applications can have either a local backend or a remote backend by which services are supported. This service model is broadly applied in mobile cloud computing since backend running in the cloud can only support Internet-based services. Therefore, access to the Internet is a fundamental condition for adopting BaaS. Mobile cloud computing is considered an efficient approach for deploying BaaS due to the advantages of mobility. Details about mobile cloud computing will be explained in Chapter 3.

Box 2.4 explains the main differences among three cloud service models, BaaS, IaaS, and PaaS.

Box 2.4: Comparison between Baas, IaaS, and PaaS.

BaaS has some similarities to both IaaS and PaaS. However, as an innovative service model, BaaS has its own unique features that distinguish itself from the two other basic service models.

BaaS focuses on the facilitation of mobile app development from a technical perspective. Cloud users can leverage BaaS services to generate new mobile applications when the required infrastructure is exactly targeted. Developers can acquire the required backend from BaaS providers after determining what backend is demanded by their new applications. In contrast, IaaS has a dissimilar concentration that intends to provide cloud customers with virtualized infrastructure. It does not usually consider the infrastructure usage from the technical perspective. Primarily, IaaS pays more attention to offering remote infrastructure that is ready to use for customers.

Next, PaaS has a similar concentration on the application development side as BaaS. Nevertheless, PaaS usually offers a complete platform for application development, from user interface to backend. This wide service scope of PaaS is different from BaaS which is solid backend only.

2.5 INFORMATION-AS-A-SERVICE

Information-as-a-Service is a cloud service model that provides cloud customers with data in an enterprise-friendly or user-friendly format, which is a service representation using a standardized schema to generate and present information efficiently. *Information* refers to the knowledge or user-readable data generated by trustworthy resources. The purpose of adopting Information-as-a-Service is to gain a way of obtaining trusted information to match business needs or individual usage purposes. This service model is an efficient approach for cloud users to rapidly acquire meaningful information from service providers without

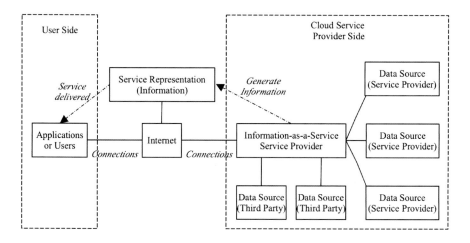

Figure 2.5 Framework of Information-as-a-Service.

physically collecting messy data, digging into a large amount of data, and wrapping up all data for results.

Figure 2.5 represents a framework for Information-as-a-Service. Information is delivered to cloud users as the service representation. Information-as-a-Service providers generate information from a data source owned by the service providers or a third party. The represented information must follow users' business rules which can be directly used by users.

One of the key values of adopting Information-as-a-Service is to enable most applications to retrieve any type of information by using an *Application Programming Interface* (API). Using an API generates an open architectural networking environment in which multiple applications can share data contents. It is an application-to-application schema supporting efficient communications and knowledge sharing. A standardized API can ensure that data from various sources is compatible information shared and used in distinct applications. Therefore, a great feature of Information-as-a-Service is helping applications to remove information sharing restrictions from the limitations caused by unstandardized APIs. A compatible operating environment enhances the usage level of data and information.

2.6 INTEGRATION-AS-A-SERVICE

Integration-as-a-Service refers to a cloud service model that provides cloud customers with system integration services. *System Integration*

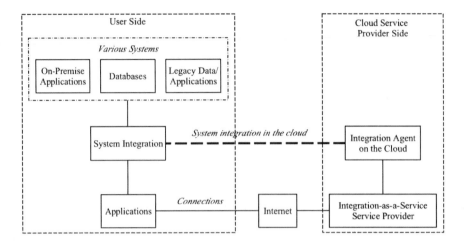

Figure 2.6   Framework of Integration-as-a-Service.

refers to a process of pulling together all subsystems running in one system. The outcome of the system integration is to enable all subsystems to work coherently and perform like a whole system. A subsystem can be a computing system, a framework, an application, or an app that runs either on-premises or remotely. For example, an enterprise's database center synchronously runs two separate systems, *Human Resource Management System* and *Client Information Management System*. Leveraging system integration can assist enterprises to strengthen both systems by duplicating and using data between systems. The value of system integration in this case is to share the functions of each system and find out the connections or relationships between employees and enterprise clients.

Figure 2.6 represents the framework of Integration-as-a-Service. When adopting Integration-as-a-Service, system integration can be done via an integration agent on the cloud. The role of the Integration-as-a-Service provider is to help enterprises integrate various systems that are either already physically being employed or new components to the whole system. Moreover, the scope of the integration is flexible. Cloud users can determine whether they want partial or total integration. A "pay-as-you-go" model allows cloud customers to purchase partial integration services instead of a total integration, which is usually costly in most situations.

There are a few techniques for delivering Integration-as-a-Service. Common approaches for on-premises systems include *Enterprise Application Integration* (EAI), *Business Process Management* (BPM), *Work*

Flow Design, App-to-App Integration (A2AI), and Web services or other networking-based techniques. For a cloud-based solution, current popular techniques are EAI, BPM, and Web services. It is important for cloud consumers to select a proper service provider whose system integration method efficiently matches the requirement of the customer's system. In this chapter, we mainly introduce EAI techniques for providing a fundamental view of Integration-as-a-Service implementations. Other mechanisms, BPM and Web services, will be presented in other chapters.

EAI is an integration framework that enables enterprises to consolidate and coordinate multiple applications on the enterprise level. The technological implementation principle is to establish a *Middleware* that coordinately connects and inter-operates all involved applications by which the usage of computing resources in enterprises is optimized. Generally speaking, all data transformation, routing, and logging can be executed in the *Middleware* no matter what interfaces the connected systems are using. To reach this goal, EAI service can be realized by a category of approaches In Integration-as-a-Service, two main approaches are the *Broker Model* (BM-EAI) and the *Bus Architecture Model* (BAM-EAI).

The Broker Model of EAI uses a central integration engine positioned in the middle of networking connections in which all inter-application functions are delivered. The central integration engine plays a broker role between applications, through which all transmissions. Besides the fundamental functionalities, some extra tasks can be done by the broker, such as monitoring and auditing. The core advantage of deploying BM-EAI is that applications can execute and communicate asynchronously since responses from recipients are not required. A crucial weakness of this model is the restrictions of the central engine. The entire system will not operate normally when the Broker's workload is excessive. A blockage of communications on the central engine can result in seriously inefficient performance of the whole system.

Moreover, BAM-EAI is the other model that is considered a novel approach to EAI. This model is designed to avoid the blockage that occurs at the Brokers. The concept *Service Bus* is introduced to EAI, which is an architectural model supporting heterogeneous applications, backends, and user interfaces. All movements that take place in the Broker are pre-defined in the *Service Bus* as working modules. Heterogeneous sub-systems can communicate with the *Service Bus* in the same way by utilizing a messaging server, such as *Java Message Service*

(JMS) and *Advanced Message Queuing Protocol* (AMQP). The messages traveling on the *Service Bus* are written in the same language format, such as *Extensible Markup Language* (XML), so that all applications attached to the *Service Bus* can communicate with each other. Compared with BM-EAI, BAM-EAI offers more reliable performance that allows service implementations to be accomplished within a few enclosed components. Figure 2.7 represents a BAM-EAI implementation model by deploying Integration-as-a-Service.

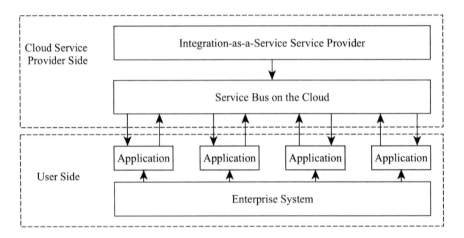

Figure 2.7 Bus architecture model of enterprise application integration using Integration-as-a-Service.

2.7   SECURITY-AS-A-SERVICE

Security-as-a-Service is a cloud service model that makes security service available to cloud customers. Current security services include both security tools and security management mechanisms, such as *Intrusion Detection Systems* (IDSs), file protectors, and encryption services. Cloud users can either outsource security operations to cloud service providers or use some security tools running on the cloud. This mechanism can assist cloud users to migrate part of risk management or full security governance into the cloud. The core value of this cloud service model is to obtain real-time updatability for security databases. For example, a cloud-based anti-virus software runs its virus database on the cloud servers. A report will be sent to the cloud-based server when cloud customers encounter a threat. The cloud server will update its

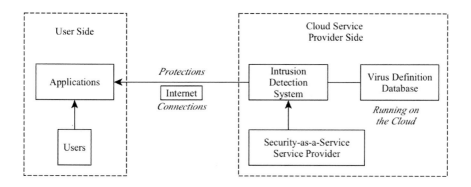

Figure 2.8 An example of a cloud-based intrusion detection system deployed by Security-as-a-Service.

database by adding the virus definition and assist cloud users to solve the vulnerability synchronously.

In many situations, security is considered an obstacle for enterprises or individual users to adopt cloud computing technologies. The main concern derives from the unknown component hidden in clouds. Lack of technical transparency often results in various restrictions when users consider their migrations of cloud-based solutions. Nevertheless, at this point, Security-as-a-Service is different from most other service models. Security-as-a-Service is on the basis of a trustworthy and reliable relationship between cloud providers and customers. Users' intentions of leveraging Security-as-a-Service are to increase their security level and strengthen secure governance capability from the trusted parties' service offerings.

Figure 2.8 represents an example of the IDS running on the cloud deployed by Security-as-a-Service. In this example, cloud servers store all virus definitions. End-users' applications are protected by the cloud-based IDS by connecting to the Internet. An Internet connection enables a cloud-based IDS to perform the same functionality as the on-premises application. The primary strength of using this solution is providing efficient updateability. New threats detected by the cloud-based IDS will be reported to the cloud servers once cloud users encounter any hostile actions. A solution response will be sent to end users when the virus definition database is updated. Nevertheless, like other cloud-based solutions, Security-as-a-Service also has a restriction due to its dependency on the Internet. A cloud-based security schema cannot execute without a connection to the Internet.

Box 2.7 illustrates the basic strategies of leveraging Security-as-a-Service.

Box 2.7: Strategies of Leveraging Security-as-a-Service.

Besides technical transparency, Security-as-a-Service users also need to know some important skills when leveraging this service model.

1. A security hierarchy will be helpful for cloud users in selecting service offerings. Security-as-a-Service users do not have to stick to only one cloud service provider. The selections of service providers can rely on the requirements of security. A proper security hierarchy can assist each cloud user in gaining a perspicuous cognition about the security essentials on different levels or in various scenarios. Based on a genuine understanding of hierarchical security demands, cloud users will have a better chance to address the security issues by choosing a cloud provider who has extreme strength in a specific security field matching the targeted needs.

2. For those cloud customers who require higher-level security, good communication between cloud providers and users is very important. Security is always a sensitive topic since it often involves the sensitive information and data. Security-as-a-Service is an outsourcing model over which cloud customers do not have a full control authority. Therefore, cloud customers must speak up about their security demands to make sure service providers can fully understand the requirements of security level and capability. Meanwhile, cloud customers should also have a basic idea about the cloud provider to ensure the selected service provider is competent for the security tasks. This means that cloud users must understand the basic technical approaches employed by the service providers.

2.8 MANAGEMENT/GOVERNANCE-AS-A-SERVICE

2.8.1 Main Concepts

Management/Governance-as-a-Service (MGaaS) offerings provided by cloud vendors help cloud customers to uniformly manage computing resources for both on-premise and remote applications. The object is to deliver end-users a managerial approach that can structurally organize, integrate, and utilize all systems or devices obtained from different sources. The main users of MGaaS are enterprises that need multiple computing resources from either local sites or diverse remote servers. The primary purpose of adopting MGaaS is to manage and improve business processes by using cloud-based solutions.

A unique characteristic of MGaaS is that this service model concerns both cloud service functionality itself and business processes, such as business rules and policies. Most MGaaS consumers intend to create business values by strengthening the operations of computing infrastructure via cloud-based approaches. For example, for a financial firm, the value creations need to match the requirements of financial business operations in a market-oriented scenario. The role of cloud-based solutions is a facilitator who can assist the financial firm to achieve its targets, including strategic, tactical, and operational goals. These targets usually depend on the real-world context and enterprises' development policies and strategies. Therefore, an effective MGaaS needs to follow users' business policies so that an efficient communication between end users and cloud vendors is a necessity.

Two core elements in MGaaS are business processes and individuals. From a practical perspective, most cloud customers need to leverage both on-premises and remote computing resources for managing business processes. Service providers need to discern cloud customers' business policies as well as business processes. Usually, this procedure is a complicated operation since many interactions and interconnections are involved within this process.

2.8.2 Mechanism

Figure 2.9 represents the framework of MGaaS. As shown in Figure 2.9, business strategies and policies can be aligned with multiple goals on various levels, such as strategic, tactical, and operational levels. Based on these goals, cloud vendors deliver services by a three-step service design process: *Design Policy*, *Define Policy*, and *Run Policy*. The outcome of the service design process is to generate a cloud-based solution that matches the requirement of value creations requested by enterprise users.

Cloud consumers need to carefully select their MGaaS service providers since this service model is based on a higher-level collabo-

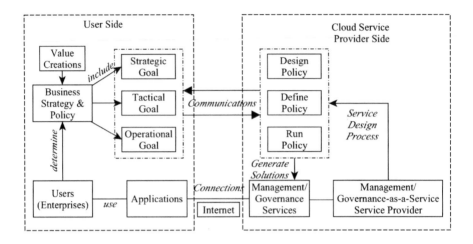

Figure 2.9 Framework of Management/ Governance-as-a-Service.

ration among all involved parties and the results of the collaboration will have a strong impact on business operations. Failure of the MGaaS adoption may result in serious consequences. Addressing security concerns, cloud vendors and customers should look at these issues from two sides, business operations and technical limitations. Three main aspects concerning this issue include *System Surveillance, Application Dependency,* and *Operation Security.*

First, cloud customers should have the authorization to surveil all activities occurring on the systems by leveraging MGaaS. All remote systems should represent the same governance performance as the on-premises systems. The chief purpose of surveilling systems is to monitor the business processes and make sure applications running in different systems can follow customer demands. Meanwhile, cloud end users do not want service providers to have the same authorization as they do. It is always a challenge to avoid service provider employees touching cloud customer data in MGaaS as well as other service models. We will discuss this topic in the succeeding chapters.

Second, leveraging MGaaS should consider whether applications are independent of one another. In a networking connected environment, there may be plenty of applications that are dependent on each other due to computing resource sharing purposes or operational requirements. However, this operational model also introduces a few restrictions. An efficient MGaaS needs to ensure that other applications can still work well when one of the applications within the system crashes. This issue is related to various aspects, such as database backup, data synchronization, database administration, data governance, system integration, and authorization management. Considering the complexity of this issue, cloud vendors should generate and provide an effective schema that can successfully identify the complexity of the hierarchy and response mechanism. Figure 2.10 represents a model of leveraging MGaaS to deal with application dependency.

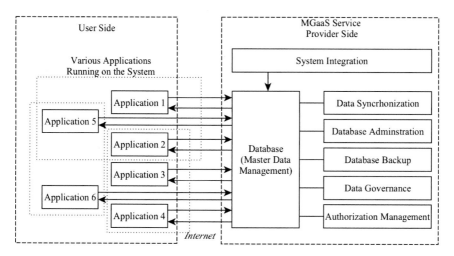

Figure 2.10 Model of using Management/Governance-as-a-Service to deal with application dependency.

As exhibited in the figure, various applications within a system may use the same database. Applications that need a shared database are grouped by dotted-line rectangles. For example, Application 1, Application 2, and Application 5 are three distinct applications that need to share one database and synchronize the shared data. The MGaaS mechanism needs to ensure Application 2 and Application 5 can successfully continue operating when Application 1 encounters problems and terminates operations improperly, which may result in various errors due to various reasons. The same solution should be available to other groups of applications illustrated by the dotted-line rectangles in the figure, such as Application 2, Application 3, and Application 4. In this case, Application 2 is involved in two groups of applications, which is a good example to emphasize the independence of applications.

One of the reasons why multiple applications need to share a database is to achieve *Master Data Management*. *Master Data Management* refers to a mechanism that supports a holistic view of data governance and management throughout the enterprise system. This is an efficient approach for enterprises that need to manage a large, complicated database that is used by numerous departments. These departments can access the database with different applications without affecting each other.

Finally, security concern is a common issue in adopting a cloud-based solution. Concerning operational security in MGaaS, cloud vendors should also consider any potential risks when cloud consumers use the MGaaS solutions. A safe mechanism needs to be generated when designing the service. Techniques for securing cloud solutions will be introduced in Chapter 8 and 9.

2.9 OTHER SPECIFIC CLOUD SERVICE MODELS

Other specific cloud service models include, but are not limited to, Information-Technology-Management-as-a-Service, Process-as-a-Service, and Testing-as-a-Service. Most current service model terms are business-related terms that are proposed by service providers. New specific cloud service models keep emerging due to real-time marketing demands and technical development. A service model is created when computing resources are converted into a type of service by cloud providers. There is a great potential that most computing resources can be represented in the cloud in the future. The next chapter introduces four main cloud computing service deployments.

2.10 SUMMARY

In this chapter, we introduced a number of specific cloud service models. These service deployments had shown a great flexibility for delivering cloud services. Understanding these specific service models could assist students in designing their own service offerings by using the available computing resources. We have also discussed the benefits and drawbacks for some models. It could help enterprises or personnels determine which cloud service was proper for their goals.

2.11 EXERCISES

Students should be able to answer the following questions after reading Chapter 2. The goals of these practices are to assist students in discerning the mechanisms of specific cloud computing service models.

1. Whose desktop is a customer using when he or she is applying a Desktop-as-a-Service solution?

2. What are benefits and drawbacks of leveraging Storage-as-a-Service?

3. Who are the main customers of Database-as-a-Service? And what are the opportunities and challenges of using Database-as-a-Service?

4. What are the differences between Storage-as-a-Service and Database-as-a-Service?

5. Company A intends to adopt cloud-based solutions to extend its database storage capacity. An important requirement of a cloud solution is that data stored in remote servers should be compatible with the existing applications and systems, such as an *Enterprise Resource Planning* system. **Question:** Which cloud service model should Company A choose Storage-as-a-Service or Database-as-a-Service? Explain your opinion in details.

6. What does *Backend* mean in BaaS? What role does BaaS play between front applications and backend supportive programs?

7. Bob is an app developer who plans to develop a new mobile app within the Android framework. However, it will be very time-consuming if Bob starts developing the app from scratch. Therefore, Bob goes to the Internet and looks for some cloud providers who can offer some kit services. Bob believes that he can save a lot of development time if he can find some cloud-based tools and use them directly in his new mobile app. **Question:** In this case, what kind of service model was Bob looking for? Provide details about the service model you chose.

8. What are the differences among BaaS, IaaS, and PaaS?

9. What are the key values of using Information-as-a-Service?

10. What does *System Integration* mean? And what are the key techniques for offering Integration-as-a-Service?

11. How could you integrate cloud customer systems if you were an Integration-as-a-Service provider?

12. What is a service scope of Security-as-a-Service? How could you make a strategy of using Security-as-a-Service if you were a *Chief Information Officer* (CIO) in your company?

13. Jasmine is using a cloud-based *Intrusion Detection System* (IDS) for protecting her laptop. Frequently, some notices from the IDS say that the virus definition has updated and synced. What do "update" and "sync" mean here? What is the process of syncing the virus definition when using a Security-as-a-Service solution?

14. What are two core elements in MGaaS?

15. Positioning yourself as a customer of MGaaS, how do you secure your cloud solution? Provide details about your plan.

16. What is the concept of cloud service deployment?

17. Picture yourself as a CIO in a company and your company has a higher-level security requirement. Now your boss intends to introduce a cloud computing solution to replace the current on-premises database. Which cloud service deployment do you think is a better solution among the four cloud service deployments mentioned in Section 1.5?

18. What are the main differences between community cloud computing and hybrid cloud computing?

19. **Design Question:** University C plans to expand their database capability by using cloud computing solutions. However, two groups of people have problems with the cloud deployment methods. One group believes that it is not secure to use a cloud-based database because the university will not fully control a cloud-based server and they are concerned about privacy leaks. For example, the *Student Registration System* has a large number of students personal information that needs to be highly protected. The other group of people argue that the university only has a very limited budget for hosting servers. Some university websites

are worth hosting. For example, there are over fifty student clubs or organizations on campus. Each of them has an individual Web page on the university's portal. Limited visiting rates result in a great computing resource waste. Therefore, using cloud-based solutions is a good choice under the university's current financial condition. Imagine yourself as a technical consultant who is hired by the university. **Your mission** is to provide University C with a detailed analysis and assist them in establishing a great investment plan using proper cloud computing service deployment.

2.12 GLOSSARY

Backend-as-a-Service also known as *Mobile Backend-as-a-Service*, is a new service model that connects applications with required backend cloud computing resources and *Application Programming Interfaces* (APIs).

Database-as-a-Service is a scalable service provisioning remote database capacity as if the database is hosted locally.

Desktop-as-a-Service is a cloud computing solution providing cloud customers with an approach of using distributed computing resources by virtualizing remote device desktop and functions.

Enterprise Application Integration is an integration framework that enables enterprises to consolidate and coordinate multiple applications on the enterprise level.

Information-as-a-Service is a cloud service model that provides cloud customers with data in an enterprise-friendly or user-friendly format, which is a service representation leveraging standardized schema to generate and present information efficiently.

Integration-as-a-Service refers to a cloud service model that provides cloud customers with system integration services.

Layered Approach is a schema that breaks the object issues into a number of layers based on criteria such as requirements, parameters, framework, or limitations.

Management/Governance-as-a-Service refers to service offerings provided by cloud vendors to help cloud customers uniformly manage computing resources for both on-premises and remote applications.

Master Data Management is a mechanism that supports a holistic view of data governance and management throughout the enterprise system.

Security-as-a-Service is a cloud service model that makes security service available to cloud customers.

Storage-as-a-Service is a cloud-based architectural solution that provides end users with data storage services, which is often considered a business model.

System Integration is a process of pulling together all subsystems running in one system.

Thin-client refers to a computer or a program that needs to be highly connected to other facilities in order to execute its computations.

CHAPTER 3

Basic Mechanisms and Principles of Mobile Cloud Computing

CONTENTS

Mobile CLOUD COMPUTING is a paradigm describing techniques, mechanisms, and applications that are used for reaching cloud computing services within a mobile environment. The definition of cloud computing was introduced in Chapter 1. In this chapter, students will understand the basic mechanism of mobile cloud computing as well as crucial techniques used in this domain.

Students will be able to answer following questions after reading this chapter:

1. What is mobile cloud computing and how can I design a high-level structure for mobile cloud computing?

2. What are three fundamental techniques supporting current mobile cloud computing applications?

3. List a few common emerging techniques used in mobile clouds and explain the mechanisms of these new techniques.

3.1 INTRODUCTION

This chapter aims to assist students in understanding the basic concepts of *Mobile Cloud Computing* (MCC). We will briefly introduce three crucial techniques in MCC, which include mobile computing, wireless networks, and cloud computing. The paradigm of MCC is basically supported by the integration of these three techniques. The content of this chapter can help students to understand the method of achieving cloud-based services by applying mobility. Partial contents of this chapter derive from our another textbook *Mobile Applications Development with Android: Technologies and Algorithms* [23].

3.1.1 Concepts

Mobile Cloud Computing (MCC) is an extension of cloud computing, which is expanded by using mobile computing and wireless networks. The concept of MCC is a mechanism using multiple technologies to achieve Web-based services by combining mobile computing, wireless networks, and cloud computing. Using this mechanism offers a variety of benefits that derive from advantages of three main technologies. For instance, using an MCC-based solution is an alternative approach for avoiding the limitations caused by the mobile computing capability, such as short battery time for mobile devices. Using MCC can also scope up the service scope of cloud computing by introducing mobile computing and wireless networks, such as mobile apps running on mobile devices that provide extensive remote storage.

From the perspective of cloud computing, there are two main workloads migrated to the remote side, including data processing and data storage. In other words, MCC-related services generally rely on two functional provisions, which include moving on-premises data processing and storage to cloud servers. Thus, the limited capabilities of mobile devices and the limited storage size of local devices can be extended when on-premises workloads are migrated to remote computing resources. In most situations, using MCC is desired to obtain higher-level service performance on mobile devices.

Furthermore, as we discussed in Chapter 1, cloud computing offers a group of service models and deployments. These cloud computing service models and deployments are also applied in MCC domains. This is because cloud computing and MCC have similar core value offerings. Both are designed to utilize distributed computing resources by applying Internet-based communications. However, there exist a few differences between cloud computing and MCC. We introduce two major differences as follows.

First, MCC and cloud computing generally have different service focuses. Both cloud computing and MCC can be implemented on wired or wireless networks. However, MCC also concentrates on delivering cloud computing services via wireless communications. Distinguishing from MCC, cloud computing emphasizes the techniques of virtualizing remote computing resources. In an MCC service model, wireless networks are considered a fundamental requirement for service deliveries. Thus, a wireless operating environment must be addressed throughout the process of MCC service designs, developments, and implementations. Recent booming development of MCC has driven great growth in intelligent mobile communications, due to the diversity of service offerings and flexible service models.

Second, MCC services mainly address utilizations of mobile computing, which include mobile devices, interfaces, and platforms. This feature leads to a distinct requirement when designing an MCC solution. For example, an MCC-based solution needs to achieve on-premises functions and also consider the extensions made by mobile computing and wireless networks. A variety of elements required for these considerations include real-time communications, wireless data synchronization, smartphone screen displays, wireless secure connections, touch-screen feature designs, etc. These wireless-related extensions are usually associated with core crucial service value creations in MCC services. Meanwhile, cloud computing services are generally broadening the service domains that cover wireless services, which is dissimilar to MCC.

Figure 3.1 shows a service flow diagram of using MCC applications. Three crucial components of the service flow include mobile computing, wireless networks, and cloud computing. Mobile devices are the carriers of cloud computing applications. Wireless networks carry all communications between mobile devices and cloud computing. Users reach the service via the service presentation interface, which is given on mobile devices. The techniques used in cloud computing will be introduced in Section 3.1.2.

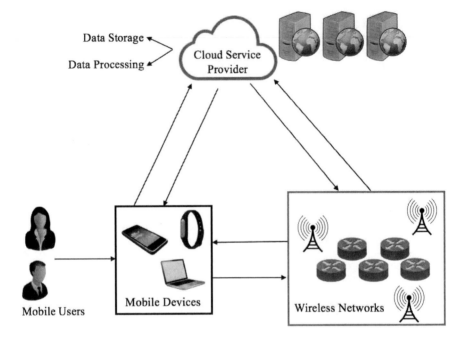

Figure 3.1 A diagram of leveraging mobile cloud computing.

3.1.2 Fundamental Components

An MCC application is usually designed to provide cloud customers with cloud-based services represented on mobile devices. To reach this goal, three key techniques are required, including *Mobile Computing*, *Mobile Internet*, and *Cloud Computing*. Combining these three aspects is a fundamental requirement for generating an MCC-based solution. We will mainly discuss mobile computing and mobile Internet in this section since we already discussed the concept of cloud computing in Chapter 1. Figure 3.2 shows the technical structure of MCC and illustrates the main techniques used in each domain. There are more techniques being implemented in practice. Students can gain a conceptual view of techniques in MCC from reading Figure 3.2. The following part of this section provides further detailed introductions of the characteristic techniques of MCC.

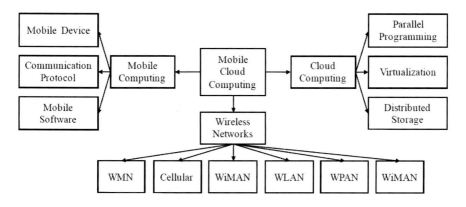

Figure 3.2 Technical structure of mobile cloud computing.

3.2 MOBILE COMPUTING

The concept of *Mobile Computing* is a series of technologies of connecting and communicating mobile devices as well as mobile platforms and service displays in networks. Mobile computing generally consists of three components, which include communication protocols, mobile devices, and mobile software. Figure 3.3 illustrates a structure mapping three key technological aspects in mobile computing. The figure also provides a few instances of each aspect.

The general concept of a *Communication Protocol* can be described as a set of rules that are designed to make sure all mobile devices within the networking communications can receive and respond to messages to each other when data are transmitted. In essence, a communication protocol needs to establish a rule to ensure the data transmissions align with the specified purpose, such as response time or message security. An example of a communication protocol is *Transmission Control Protocol* (TCP) and *Internet Protocol* (IP), which is generally known as a *TCP/IP*. In most situations, TCP/IP is considered a reliable and secure protocol for data transmissions. Meanwhile, another communication protocol, *Connectionless User Datagram Protocol* (UDP), is generally considered an efficient communication protocol but can only offer a lower-level security standard.

The mobile device is another side of mobile computing. The term *Mobile Device* includes those portable/wearable devices that can carry wireless communications for executing mobile applications. A few ex-

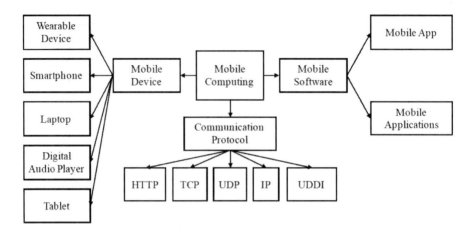

Figure 3.3 Three key aspects of mobile computing and examples.

amples of mobile devices include laptops, tablets, smart phones, smart watches, digital audio players, portable gaming systems, and other wearable devices. As a service presentation platform, mobile devices are an interface for mobile end users to acquire Internet-based services.

The concept of *Mobile Software* refers to those applications/apps that are executed on mobile devices for mobile service offerings. In this definition, we emphasize that the meanings of mobile applications and mobile apps are different, even though both can describe a service representation program designed for mobile devices. A mobile application needs to be executed/run directly on the *Operating System* (OS). But a mobile app needs to run within the framework. A common framework for mobile apps is Android. The purpose of using the framework is to shorten the app's development time.

3.3 WIRELESS NETWORKS

Wireless Networks, also known as *Wireless Networking Technology* (WNT), or *Mobile Internet*, refers to a set of networking technologies designed for connecting communicators over wireless networks and supporting mobile software. The *Communicator* refers to a *Network Node* in wireless networks, which essentially means various workload processing locations and infrastructures. Any connectable and communieable points with functions or purposes in wireless networks can be network nodes. Activities that occur at network nodes need to follow the networking rules required by communication protocols. For example, a network node can configure its succeeding activities in a routing protocol. Two basic types of network nodes are connection and endpoint nodes, which represent two kinds of data transmissions.

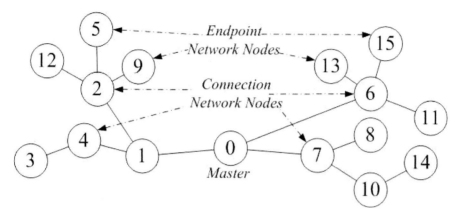

Figure 3.4 An example of network nodes in a tree topology for a telecommunications network.

Figure 3.4 exhibits an example of essential network nodes deployed in a *Tree Topology* for a telecommunications network. A tree topology is a tree-shaped topology that displays the dispositions and configurations of computing elements for producing a network. Connection network nodes take responsibility for bridging communicators and connecting communications. In the middle of links, connection network nodes can also split or converge the data transmissions and allocate the transmissions onto new paths. In Figure 3.4, nodes 2, 4, 6, and 7 are examples of connection network nodes. Endpoint nodes are terminal points of the wireless networking system, in which the information or data are abstracted and represented in a machine-friendly format. A terminal point is a networking spot in which communications launch or end. In a wireless network, an endpoint network node is a mobile device in most situations. For instance, nodes 5, 9, 13, and 15 are endpoint network nodes in Figure 3.4.

Currently, there exist a few wireless network types designed for delivering different services or achieving different purposes. Some common types include *Wireless Personal Area Networks* (WPANs), *Wireless Local Area Networks* (WLANs), *Wireless Mesh Networks* (WMNs),

Table 3.1 Comparison of Current Various Types of Wireless Networks.

Type	Performance	Applications	Coverage Capability
WPANs	Moderate	Personal use	Small
WLANs	High	Community or organization use; connecting mobile devices via an access point connected to the Internet (Wi-Fi)	Within an area
WMNs	High	Using radio nodes within a mesh topology to connect mesh clients that are mobile devices	Global
WiMANs	High	Cover larger areas than WLANS; the wireless service is offered and operated by an organization in most situations	Within a city
WWANs	Low	Mobile access to wireless networks	Global
Cellular	Low	Mobile access to wireless networks by using cell-style distributed radio networks	Global

Wireless Metropolitan Area Networks (WiMANs), *Wireless Wide Area Networks* (WWANs), and *Cellular Networks* (Cellular). Table 3.1 displays the main features of these wireless networks. The communication performance is usually associated with the service coverage area, network bandwidth, or signal station deployment.

Note: The *Wireless Mesh Topology* is a wireless networking type that interconnects all mobile devices.

Considering the deployment, there are a variety of communication methods supporting wireless networks. Two essential methods are peer-to-peer and point-to-point. The *Point-to-Point* communication method connects network nodes. The *Peer-to-Peer* communication method establishes a network connecting network nodes that can act as networking clients and servers. This type of communication method is usually deployed in a *Decentralized Network System* (DeNS). In a DeNS, all network nodes only work on the local operations such that all net-

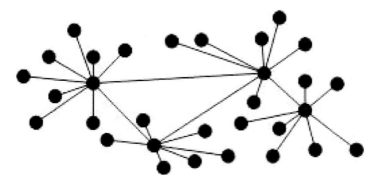

Figure 3.5 A decentralized network system.

Figure 3.6 A centralized network system.

work nodes have equal responsibilities to the network system. Figure 3.5 represents a decentralized network system. The whole system is a distributed-shape network involving a few network nodes. Another common method is the *Centralized Network System* (CNS), which consists of a master node and endpoint nodes. Figure 3.6 shows an example of a centralized networking structure. A CNS is usually employed for a smaller user community for group work in a small workplace. Box 3.3 describes main difference between two resemblant concepts: WLANS and Wi-Fi.

Box 3.3: Comparison between Wireless Local Area Networks and Wireless Fidelity

The real meaning of WLAN is different from *Wireless Fidelity* (Wi-Fi). Wi-Fi is a typical application of WLANs and Figure 3.7 represents the structure of Wi-Fi. In general, Wi-Fi products need to follow the 802.11 wireless protocol family in WLANs. And the network source of Wi-Fi can be either wired or wireless Internet sources, which is illustrated in Figure 3.7. Meanwhile, WLANs users access the wireless networks via radio connections.

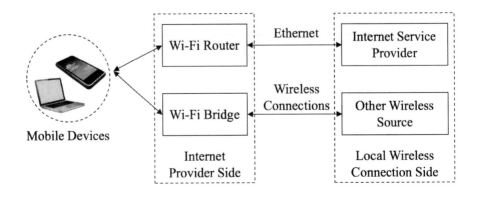

Figure 3.7 A Wi-Fi structure.

Due to the space limitations, rest of this section only briefly introduces WWANs and cellular networks, which are two typical network deployments for mobile computing. WWAN is a type of wireless network that is deployed in a large geographic area by setting a great number of radio signal cells for both mobile and local wireless users. The major benefit of using WWAN is that fast-moving objects can be well served in the networks, which is much better than WLAN. A few common communication technologies include *Global System for Mobile Communications* (GSM), *Code Division Multiple Access* (CDMA), and *Worldwide Interoperability for Microwave Access* (WiMAX). Table 3.2 provides a comparison between WiMAX and Wi-Fi.

Next, a *Cellular Network* is a type of wireless mobile network that consists of a set of radio base stations that offers networking services. Each radio base station is responsible for a specific coverage area that is called a *Cell*. To avoid interference from different cells, adjoining cells do not share the same radio frequency. Figure 3.8 shows a basic cellular network using three frequencies. A group of adjoining cells can be called a *Cluster*. The frequency can be reused when all adjoining cells use distinct frequencies.

3.4 MAIN TECHNIQUES IN CLOUD COMPUTING

There are many techniques used in cloud computing. We list three common and crucial techniques applied in contemporary cloud systems:

Table 3.2 Comparison between WiMAX and Wi-Fi.

	WiMAX	Wi-Fi
Standard	IEEE 802.16	IEEE 802.11 standard family
Coverage	Larger area, up to 40 miles coverage per WiMAX antenna	Smaller area, usually up to a few hundred feet
Bandwidth	Adjustable bandwidth range, from 1.25 to 20 MHz	Fixed bandwidth, 20 or 25 MHz
Mobility	Available for mobile users	Not supporting fast moving mobile users
User	A larger scalability	Limited number of users

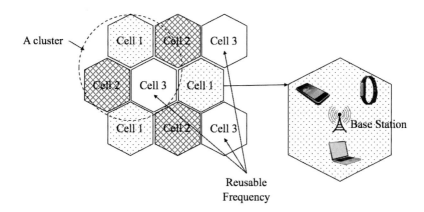

Figure 3.8 An example of the cellular network structure.

virtualization, parallel models, and distributed storage. More detailed information can refer to our another textbook [23].

Virtualization *Virtualization* is a type of techniques that offers the service of the computing resource by virtualizing the infrastructure or applications in order to achieve the service representation on users' demands. In cloud computing, a VM is a good example for applying virtualization techniques to provide services. A variety of services are

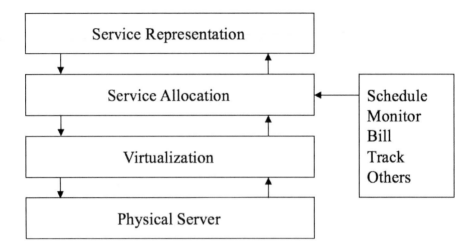

Figure 3.9 An example of using a virtual machine to deliver cloud services.

available by using a VM, such as infrastructure, platform, applications, or security and maintenance, which covers a broad scope of computing, from apps to operating systems. Some common advantages of using VMs include saving cost, reducing energy consumption, and simplifying maintenance. Meanwhile, using a VM-based technique can assist the cloud vendors in offering computing services to multiple users from the same resource pool. Individual cloud user/customer can receive a full service even though each user/customer is accessing the resource shared by other users/customers. This technique can enable cloud vendors to support a large number of users with a lower-level resource supply, which makes sense in saving financial budget and reducing energy cost from the perspective of the cloud [3, 5, 24].

Figure 3.9 shows an example for using a VM to deliver cloud services. In the figure, the service representation layer is the interface in which cloud users and service providers interact. Individual cloud user accesses the service via the interface that acts as a specific service for each user. At the service allocation layer, different service requests are forwarded to the corresponding service providers. Meanwhile, the virtualization layer shows the independent computing resource offerings to users. Finally, the physical server layer is where the workload is done on the remote server.

Parallel Model   *Parallel Model* (PM) is a category of technology that aims to solve the parallel tasks. The technology is widely applied in cloud computing due to the demand of task distributions. The mechanism principle of a PM is concurrently manipulate a few operations by using separate remote cloud servers. Cloud vendors can apply some optimization techniques to increase the efficiency of the operation. We will discuss the optimization of the parallel model in Chapter 6.

Distributed Storage   In cloud computing, *Distributed Storage* is a popular service deployment for increasing service level of data storage [20, 21]. Storage system designers need to consider the priority when they plan to deploy a distributed storage in cloud computing. For example, a few dimensions on performance increases include data security, data reliability, data storage size enhancement, and infrastructure credibility. The term *Mass Distributed Storage* (MDS) is used to describe a cloud data storage technique that combines both large size storage (*Cloud Mass Storage*) and distributed deployment. The application of MDS can address certain issues when the system is designed, such as increasing data reliability or improving infrastructure credibility. The main concerns of the distributed storage in cloud computing include: (1) whether the data are retrievable from different storage sources, and (2) data security during the transmissions and on the cloud server. We will discuss this issue in Chapters 8 and 9.

3.5 SUMMARY

In this chapter, we briefly introduce a few basic mechanisms used in MCC as well as main techniques applied in current cloud solutions. Three basic technologies of MCC include cloud computing, mobile computing, and wireless networks. We have discussed their concepts and main components. Moreover, we also talk about the main techniques used in cloud computing, which are virtualization, parallel model, and distributed storage. More information about these technologies will be introduced in the later chapters.

3.6 EXERCISES

1. What are the main workload migrations when adopting mobile cloud computing?

2. What are the differences between cloud computing and mobile cloud computing?

3. What are three key components in mobile cloud computing?

4. What does *Mobile Computing* mean? What is a communications protocol?

5. Is an app the same as an application? Explain your answer.

6. Explain the following concepts and their relationships: mobile Internet, communicators, and network nodes.

7. List at least five common wireless network types.

8. Explain the relationship between Wireless LANs and Wi-Fi.

9. **Design Question:** Bob's company intends to establish a local wireless network for business usage purposes. The aim is to enable all mobile devices running in the company to have access to the Internet. The company owns three offices sized 100 square feet. Bob hires four employees, so the wireless network needs to support at least five users including Bob himself. Some extra users are required for the company's guests. Picture yourself as a technician professional who is assisting Bob in establishing the wireless network. Your mission is to give Bob advice regarding the wireless network type that has fits the company's needs and explain why. A comparison is required so that Bob will know what wireless network types are available and why the suggested type is good for his business.

10. What are key components of a cellular network? What is a cluster in a cellular network? Use your own words to explain how a cellular network operates.

11. What are the main techniques being used in current mobile cloud computing solutions?

12. What is a process for implementing virtual machine technology in mobile cloud computing?

13. In most situations, using a virtual machine is considered a secure approach for representing cloud services. Please explain the theoretical basis.

14. What is the *Parallel Programming Model* technology and why do cloud providers need to use this technology?

15. What is the operational principle of *Mass Distributed Storage* in mobile cloud computing?

16. **Discussion Question:** What are the differences between cloud computing architecture and mobile cloud computing architecture? Think about this question from a technical perspective and figure out whether these two architectures are using the same technologies.

3.7 GLOSSARY

Application Distribution Platform is an interface for provisioning software purchases and downloads, which is owned and operated by OS providers.

Centralized Network System is a network model consisting of a master node and endpoint nodes, which is suitable for a small group of users who work together in a limited space.

Channel-Access Schema is a mechanism that enables different data streams or radio signals to go through or share the same communication channel.

Cloud Distributed Storage is a means that stores data on multiple remote databases or nodes.

Cloud Mass Storage refers to a schema that stores users' large data in remote servers and ensures the data are readable across various interfaces.

Code Division Multiple Access (CDMA) is another approach for mobile communications that can transmit multiple digital signals by connecting multiple terminals through the same transmission medium.

Communications Protocol is a set of rules that ensure all mobile devices involved in the communication can understand each other when data are transmitted.

Decentralized Network System is a distributed network model in which each networking node only works on local operations and has equal responsibility to the network.

Function Parallelisms also known as *Task or Control Parallelism*, is a method of parallel computing between multiple processors that distributes processes or threads between parallel computing nodes.

Global System for Mobile Communications (GSM) is a global standard for operating mobile communications, which was developed by the European Telecommunications Standards Institute.

Interacting Processes also known as *Processes Interactions*, is an approach used for increasing operating efficiency by interacting parallel processes.

Mass Distributed Storage (MDS) is a novel technology that combines two techniques including mass storage and distributed storage in order to apply multiple storage servers as well as enhance data reliability and infrastructure credibility.

Message Passing is a term describing a solution of exchanging data between processes, which is achieved by passing messages.

Message (in Message Passing) is a type of communication content between parallel programs, which can be generated in a number of forms, such as functions, signals, and some types of data packets.

Mobile App refers to those computer programs running on mobile devices that are developed and executed within a framework, such as Android.

Mobile Cloud Computing is an implementation pattern that leverages cloud-based solutions by using mobile technologies and devices.

Mobile Cloud Computing Architecture is a structure of designing, developing, and implementing mobile cloud computing solutions.

Mobile Computing is a group of technologies for building the links and communications between mobile devices.

Mobile Internet is a set of advanced networking technologies that activate inter-connectivity between communicators over wireless networks by which mobile software is supported.

Mobile Software refers to those applications running on mobile devices that are used to represent mobile services.

Network Nodes refers to multiple types of processing locations in various digital infrastructures.

Parallel Programming Model (PPM) is a technology used for solving concurrent tasks based on a cloud-based platform.

Peer-to-Peer (P2P) is an approach of building a network by connecting networking nodes that are both networking clients and servers.

Point-to-Point is a communication method supporting wireless connections between networking nodes or endpoints.

Problem Decomposition refers to a methodology that formulates parallel programs.

Shared Memory is an interacting process method that supports data passing between programs based on a shared global address space in which all parallel programs can read and write data asynchronously.

Spread-Spectrum is an approach that extends bandwidth capacity by varying the frequency of the transmitted signals.

Tree Topology is a tree-shaped topology that displays the dispositions and configurations of computing elements for producing a network.

Virtualization is a mechanism that virtualizes object computing resources to represent them in a service-based manner by various service levels.

Virtual Machine is an emulating technique that is used to distribute and virtualize multiple remote computing resources and present the computing acquisitions to end users in a service mode.

Wireless Local Area Network is a wireless deployment method for a small-range area usage purpose through a high-frequency radio connection.

Wireless Mesh Topology is a wireless networking type that interconnects all mobile devices.

Wireless Station in WLANS refers to any equipment connecting to the wireless medium within a wireless network.

Wireless Wide Area Network (WWANS) also known as *Wireless WAN* or *Wireless Broadband,* is a large geographic-usage wireless network in which a great number of cells transmit radio signals to both mobile and on-premises devices.

CHAPTER 4

Mobile Cloud Computing Architecture Design, Key Techniques, and Challenges

CONTENTS

Architecture

DESIGN OF MOBILE CLOUDS is a fundamental dimension of installing mobile cloud computing. This dimension consists of a few cubes that form the knowledge construct in the field. The main cubes include architecture design, key techniques, and main challenges identified in the mobile cloud domain. This chapter provides students with a holistic view of mobile clouds by offering knowledge of mobile cloud architecture design techniques and clarifying the main challenge issues. A group of cloud models are compared to assist students to further discern the concept of cloud computing. Students will be able to answer the following questions after finishing this chapter:

1. What are the crucial layers in cloud computing architecture?

2. What are the differences between cloud computing and mobile cloud architecture?

3. What are the management strategies for distributed clouds?

4. What is *Intelligent Workload Factoring* (IWF) and how does it operate?

5. What are the cloud-related security issues/concerns? And solutions?

6. Describe cloud-based privacy protection techniques.

4.1 INTRODUCTION

As discussed in Chapters 1 and 3, cloud computing technologies have turned into a powerful Internet-based paradigm that performs scalable services or creates new service contents. Complex computing can be implemented in a cloud-based scenario. Introducing mobility to cloud computing further extends the service scope by deploying various architecture strategies. In this chapter, we will consider the implementations or mechanisms of different cloud architecture deployments and discuss both the advantages and disadvantages of each strategy.

This chapter is organized in the following order. Section 4.2 introduces the main layers of the typical cloud computing architecture. Section 4.3 introduces the architecture of mobile cloud computing and identifies the main differences between cloud computing architecture and mobile cloud architecture. Section 4.4 discusses a few crucial management strategies for distributed clouds. Section 4.5 presents the optimization mechanisms of hybrid cloud computing. We demonstrate security issues and solutions in cloud systems in Section 4.6.

4.2 CLOUD COMPUTING ARCHITECTURE

As discussed in Chapter 1, cloud computing is a concept integrating a variety of computing technologies. This characteristic gives cloud computing inherent features of multiple techniques. Being aware of this fact can assist the cloud computing architecture designer to increase the performance without adding more computing facility. Figure 4.1 illustrates a high-level cloud computing architecture that has four layers: User Interface, Cloud Resource Manager/Scheduler, Virtual Machine, and Physical Machine layers [3, 25]. These four layers are basic components of cloud computing.

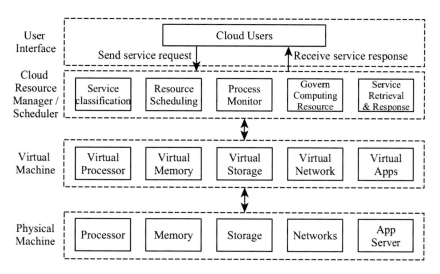

Figure 4.1 A high-level cloud computing architecture with four layers.

According to this cloud computing architecture, cloud users send service requests via the Cloud Resource Manager/Scheduler layer. The service deliveries are based on the implementation of virtual machines, by which the service contents are represented to users. The virtualized applications or hardware can be executed by either single or distributed multiple physical machines. The service responses are sent back to cloud users via virtual machines while the service is done by physical machines. This architecture can assist us to further understand the mechanism of mobile cloud computing architecture.

4.3 ARCHITECTURE OF MOBILE CLOUD COMPUTING

4.3.1 Overview

Mobile Cloud Computing Architecture is a structure of designing, developing, and implementing mobile cloud computing solutions. The architecture consists of a set of technologies that are introduced and evaluated in this chapter. Deploying an MCC-based solution requires three critical supportive aspects, including mobile computing, mobile Internet, and cloud computing [26]. Each aspect is supported by a number of core techniques. Figure 4.2 illustrates the architecture of MCC, which shows the approach of deploying an MCC solution.

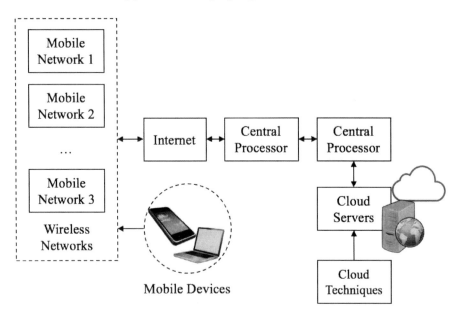

Figure 4.2 Architecture of mobile cloud computing.

According to the figure, mobile cloud users gain mobile cloud services by accessing mobile computing. Mobile devices and Internet communications protocols are two prerequisites. Currently, mobile users have a number of choices of wireless networks that can be used for various demands and mobile equipment [27]. Once users' service requests are received, the requests will go through central processors and the mobile network server for allocating services and connecting to cloud servers. Three key technologies support the implementation of cloud computing: *Virtualization*, the *Parallel Programming Model*, and *Mass Distributed Storage*. The service responses will be sent back to mobile users when the physical machines running on the remote servers finish data processing. Mobile cloud users obtain services via a virtual machine that provides cloud users with a service representation.

4.3.2   Hybrid Cloud Computing Architecture

Contemporarily, enterprises need to use cloud service deployments in order to balance the benefits and drawbacks of each service deployment [15]. High performance is one of the desired focuses for many current enterprises that intend to leverage cloud services [28, 29, 30]. Companies may consider the optimal *Total Cost of Ownership* (TCO) in order to maximize the value of using cloud-based solutions. In the cloud field, TCO is a term estimating the direct and indirect consumption of a system investment. Considering this factor will impact the strategies of service deployment selections.

Understanding this fact can help us be aware of the motivations of using hybrid cloud computing as well as the key issues in designing hybrid cloud computing architecture. There are mainly two advantages of deploying hybrid cloud computing. First, hybrid cloud computing can provide a higher-level scalable service due to the flexible options of several service deployments. The other major benefit of using hybrid clouds is that this approach may maximize the return on investment because of a larger pool of computing resources and adjustable resource configurations.

Moreover, designing a hybrid cloud computing architecture needs a selection of the dimension(s), such as workload and execution time. This needs to be determined by service providers or cloud users. Figure 4.3 represents a model example of hybrid cloud computing considering dynamic workload situations [25, 31]. This architecture can enable users to run their applications on hybrid infrastructure.

As shown in Figure 4.3, the basic mechanism of this model divides the entire task into two parts, which includes task A and task B. The task assignments will depend on both input requests and server features. When the input service requests are received, the first step

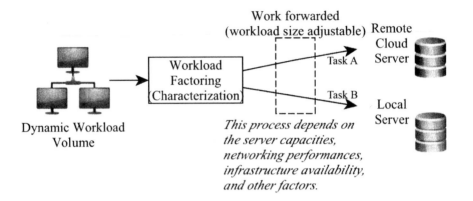

Figure 4.3 Example of the hybrid cloud computing model considering the dynamic workload dimension.

characterizes the workload volume. The data processing tasks will be forwarded to either remote cloud servers or local cloud servers based on the workload volume. The following gives a very simple example to help you understand the basic operating principle of the hybrid cloud computing model.

Assume that there is an input data processing request consisting of 10 data requests and there are two servers available, including a remote cloud server and a local private cloud server. A remote cloud server needs more processing time but costs less than the local server. The local server can process the data faster with much higher costs than the remote. All data requires the same time period and cost on the same server. Table 4.1 provides the assumptions of the required time and cost on both the remote cloud server and local server. The formula of the time consumption is *the number of data × Time Per Data*. The cost can be obtained by *The number of data × Cost Per Data*. For instance, the required time is 32 when we assign 4 data to the remote cloud server and 6 data to the local server, which is from $(4 \times 5 + 6 \times 2)$. Correspondingly, using the same assignment, the cost is 54, which is from $(4 \times 3 + 6 \times 7)$.

Our goal is to assign these 10 data to these two servers to satisfy the service request. We configure four assignment settings in order to figure out the performance differences by a comparison between task assignments. The details of the four settings are given as follows:

Table 4.1 Example of Data Consumption on Different Cloud Servers.

	Time Per Data	Cost Per Data
Remote Cloud Server	5	3
Local Cloud Server	2	7

- **Setting 1:** Assign all 10 data to the remote cloud server.

- **Setting 2:** Assign all 10 data to the local server.

- **Setting 3:** Assign 5 data to the remote cloud server and 5 data to the local server.

- **Setting 4:** The assignment that can gain the minimum cost under timing constraint of 30.

After computations, we have the results that are shown in Figure 4.4. There are many ways to solve this problem. We give one solution as follows. Assume there are x data assigned to the remote cloud server so the number of the data assigned to the local server is $(10-x)$. According to the formula given above, the required time (T) can be determined by $[5 \times x + 2 \times (10 - x)]$; the cost value (C) can be determined by $[3 \times x + 7 \times (10 - x)]$. Therefore, we have the following equation to calculate the time and the cost.

$$\begin{cases} T = 3x + 20 \\ C = -4x + 70 \end{cases}$$

Open Discussion Question: Think about the results of Setting 4 shown in Figure 4.4. Give your solution achieve the result and have a discussion about whether this result is an optimal solution based on the conditions and requirements.

4.4 CRUCIAL MANAGEMENT STRATEGIES FOR DISTRIBUTED CLOUDS

Figure 4.5 represents an operational process architecture of resource management that allows fault-tolerance cloud computing. The fundamental operating principle uses abstract computational resources by

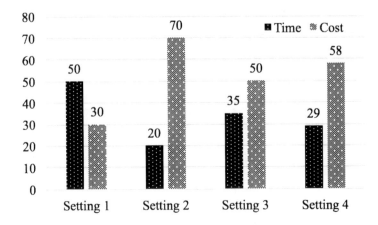

Figure 4.4 Time and cost performance in different settings for the given example.

which the function blocks are formed. The management operations are based on the information of function blocks for scheduling or organizing task allocations. In this book, the *Function Block* (FCB) is defined as a certain function achieved by a specific cloud resource in distributed cloud computing. The *Cloud Resource Manager* is a layer in the cloud computing resource management architecture, in which a series of resource operations are accomplished, including resource information retrievals, task assignments, computation process displays, communication interconnections, and operation surveillances. This layer can be either physical infrastructure or software-based solutions using virtual operating interfaces. There are a few basic subprocesses in resource management.

1. Identify FCB in order to discern each cloud service provider's capacity and service offerings.

2. Need to obtain information about a series of operations at various states, such as start state, end state, query state, assignment state, and processing state.

3. The computation process time needs to be displayed to the cloud users. The information can be retrieved from cloud resource managers. The required and used cloud computing resources are viewable for the purpose of the scalable services.

4. The cloud resource manager configures and maps computing resources by retrieving information from multiple cloud providers. The service abstracts need to be represented to cloud users at the business layer.

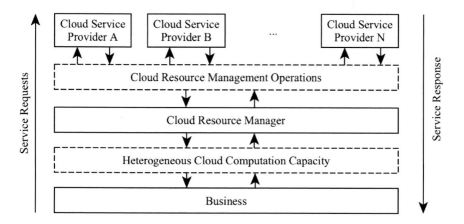

Figure 4.5 Architecture for resource management allowing fault-tolerant cloud computing.

Normally, the service representations are duplicatable. Once the cloud resources are mapped by a certain cloud service provider, the cloud resource manager can gain the computing resource mapping information from prior mapping schemes. Current generic representations of computation resources can be used to synchronously illustrate available cloud computing resources offered by various cloud providers. The computing information is a crucial aspect in optimizing cloud computing because an effective task assignment can enable high performance by implementing parallel computing and avoid idle cloud facilities [11, 32]. The optimization algorithms will be introduced in the succeeding chapters.

4.4.1 Hybrid Cloud Resource Manager

A *Hybrid Cloud Resource Manager* is a typical cloud resource manager that organizes, interconnects, monitors, and facilitates operations for a group of different cloud providers. These cloud providers usually offer similar or relevant cloud services so that the resource allocation is the critical issue in hybrid cloud computing optimizations. A general goal of a high-performance hybrid cloud resource manager is to effectively organize the available computing resources from multiple cloud providers for achieving a higher-level working efficiency or minimum resource costs. In this section, we use a contemporary real case of applying a hybrid cloud resource manager to briefly introduce its mechanism:

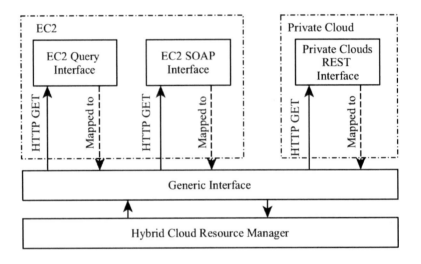

Figure 4.6 High-level manipulative structure of Amazon Elastic Compute Cloud. Both private cloud and EC2 interfaces are supported.

A contemporary instance of applying a hybrid cloud resource manager is given by Amazon *Elastic Compute Cloud* (EC2) [33]. In this instance [34], on one hand, the tandem of Amazon EC2 is governed by Query and a *Simple Object Access Protocol* (SOAP) interface; on the other hand, the private cloud is conducted by the *Representational State Transfer* (REST) interface. For successfully delivering cloud services, some work is done by the hybrid resource manager. Figure 4.6 represents a high-level manipulative structure of Amazon EC2 supporting both EC2 and private cloud interfaces. The detailed mechanism is described in this section, too.

First, the EC2 Query interface applies a query string to the *Uniform Resource Locator* (URL) by which the cloud resource manager is deployed. Next, as the cloud service provider, Amazon produces and provisions the list of pre-defined parameters as well as the corresponding values, which are all involved in the query string provided by the cloud resource manager. When the cloud service is requested, the cloud resource manager sends the query strings to a URL via HTTP GET messages. The management operations will be enabled once this operation is triggered. In this case, the interface of EC2 is mapped to a generic interface so that the cloud resource manager is able to operate. The *Generic Interface* is the manipulative interface operated by the cloud resource manager.

Moreover, the EC2 SOAP interface has a similar operation to the Query interface. The cloud resource manager also sends out an HTTP GET message to a specific URL provisioned by the cloud provider, by which the cloud management operations can be executed. The difference between these two operations is that the actual parameters required by the operations are varied.

Furthermore, distinguished from two operations above, a private cloud REST interface needs to assign a global identifier [4] to each local resource. The identifier used by EC2 is a *Uniform Resource Identifier* (URI), which is a string of characters identifying the resource name by which the interactions of the resource on the networks are enabled. The manipulations of local resources are also based on HTTP. The interface mapping method is the same as the methods used by the SOAP and Query interfaces. This means the private cloud interface needs to be mapped to the cloud resource manager's generic interface as well. The cloud resource manager can administer and organize both private cloud and EC2 interfaces as long as each interface is mapped by the manager's generic interface.

In summary, the case of Amazon EC2 provides us with a great example of operating hybrid cloud computing by the resource manager. The cloud resource manager supervises the cloud systems by creating a generic interface. The following section will explain how the cloud resource manager can improve cloud performance.

4.4.2 Manipulations of the Cloud Resource Manager

Question: *How can hybrid cloud resource manager improve cloud computing performance?*

The hybrid cloud resource manager is able to increase cloud system's performance because of different features owned by each interface. The following is a quick review of the interface pros and cons to discern their performance diversity. The pros and cons of SOAP and REST are introduced in Chapter 10.

The diverse characteristics of these interfaces result in various performances. Figure 4.7 represents an example comparing average response time between Query and SOAP interfaces. We examined EC2 by sending out 1000 identical management operation commands via both Query and SOAP interfaces. Each operation command was labeled by timestamps showing the sending and returning time. The average response time of the Query interface is 508 ms, which is remarkably shorter than the SOAP interface. It implies that selecting the Query interface can reduce the response time, which might be critical for some businesses requiring responsive performance. Understanding this mechanism can assist cloud resource manager designers to maximize the cloud system's performance and increase *Quality of Service* (QoS) without adding infrastructure.

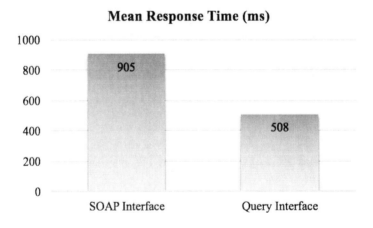

Figure 4.7 A comparison of response time between Query and SOAP interfaces.

4.5 OPTIMIZATION MECHANISMS OF HYBRID CLOUD COMPUTING

There are many optimization mechanisms proposed for improving performances of hybrid cloud computing. The significance of considering optimizations in cloud computing mainly concerns cloud architecture designers or service providers who aim to offer a higher level cloud service. In this section, we briefly introduce a few optimization methods that have been explored in prior research. Some of the optimization methods will be discussed in detail in later chapters.

First, the task assignments for scheduling optimizations in IaaS have been explored by recent researches. Most mechanisms were based on being aware of the workload across heterogeneous clouds or data centers. Some optimized task schedules were designed to manage data processing workload within a heterogeneous cloud environment. One optimization approach was proposed to schedule preemptable tasks on IaaS cloud systems. This approach uses a dynamic cloud min-min scheduling algorithm. Cloud managers store all parameters of the task executions for the references of the optimizations. The updates will be made according to the feedbacks of the task executions. The task assignments depend on the parameters displaying the earliest resource available time and the earliest finish time, by which short execution time can be ensured. In a later chapter, we will teach this approach in detail.

4.6 SECURITY CHALLENGES AND SOLUTIONS IN MO-BILE CLOUDS

As an emerging technology, mobile clouds also face a variety of threats from both operational and technical aspects. We focus on the technical perspective in this section and provide an overview of security issues in mobile cloud systems.

4.6.1 Main Challenges in Mobile Clouds

Open Discussion Question: Are there any security issues in mobile clouds that are different from traditional security concerns? If yes, what has made the security issues different?

In this section, we discuss a number of main challenges in the current mobile cloud field, such as operational abuse on the cloud side, vulnerability caused by unencrypted data, lack of trust and the multitenancy issue, malicious intrusions, and loss of controls on the customer side.

Operational Abuse on the Cloud Side: Despite many advantages of using cloud-based solutions, many enterprises are still concerned about their data security or privacy threats when they use cloud services. The threats can be caused by various aspects, even though most cloud vendors promise that they can provide the same security level as on-premises solutions. Besides traditional security concerns, such as networking vulnerabilities or malicious attacks, mobile cloud computing also needs to deal with weaknesses caused by the cloud models. The privacy issue is always a great concern for most cloud users because the data are hosted, processed, or stored in the cloud, and users do not even know where the server is. Employees working for cloud service providers can easily have access to the data that contain sensitive information or critical system data. This has resulted in wide concern, which became an obstacle in implementing cloud technologies.

Vulnerability Caused by Non-Encryption: In the non-cloud operating environment, data owners can encrypt their data in order to prevent any unexpected data users from abuses. However, in the cloud-based context, most encryption approaches cannot be used since encrypted data are difficult to execute for operations. For data processing purposes, the cloud-side memory needs to have the original data. For instance, as we mentioned in the prior section, Amazon EC2 is a good option for creating cloud services but the data are not secure.

Trust Issues and Multi-Tenancy in Cloud Computing: The trust issue in cloud computing usually considers cloud service providers as the trust party unless the research problem identifies that the providers are also potential threats in the service processes. In this subsection, we mainly talk about the common situations of trust issues related to multi-tenancy rather than targeting the trust threats caused by the cloud providers. *Multi-tenancy* is the service architecture that supports multiple customers sharing or utilizing a certain single software application. The term *Tenancy* here refers to cloud customers who may have some level of authentication for customizing the service scope. This is a broadly accepted cloud service model using virtualization techniques, which generally support SaaS.

The major challenge of trust issues between tenants is mainly caused by sharing resources via virtual machines. The concern derives from the opposing demands between tenants who share a pool of computing resources. Separating isolated tenants from a shared resource pool is a challenging issue. The security risk is that multiple independent users share the same physical infrastructure when tenanting the same applications. Therefore, the adversary will be located in the same physical machine as other cloud customers. Attackers will have legitimate status unless they are detected.

Intrusions in Wireless Networks: In wireless networks, there are a variety of vulnerabilities during wireless data transmission in which the intrusions usually occur. Adversaries enter the networks in a variety of manners, such as jamming, monitoring, DoS, or man-in-the-middle. Most threats caused by wireless connections take place when the wireless networks are deployed for cloud applications.

Customers' Loss of Control: This issue is usually relevant to privacy issues in the cloud. When cloud users apply cloud services, identity management is operated by the cloud providers. The security mechanism is designed and executed on the cloud side such that cloud customers have very limited authentications to configure the access control rules and policies. The services' availability and security highly rely on the security mechanism provided by the cloud providers. Therefore, cloud customers cannot fully control their data security.

4.6.2 Overview of Security Solutions

To solve the security problems, a variety of approaches have been proposed or developed in prior research. We list a number of common approaches addressing security problem solving in cloud computing. Some of the solutions will be introduced in later chapters. Currently, common approaches for securing cloud computing include:

Access Control Methods: This is an approach securing computer systems by defining the access rules to determine whether the user's access request is granted or rejected for the purpose of system protection. A few access control focuses include authorization, authentication, access approval, and audition. Some common access control methods include *Multi-tenancy Based Access*

Control (MBAC), *Attribute-Based Access Control* (ABAC), *Role-Based Access Control* (RBAC), and *Identity-Based Authentication* (IBA).

Data Encryptions: In cloud computing cryptography, data encryption refers to the process of encoding and decoding data from information leakage. Considering the operating environment in cloud systems, encryption methods need to match the requirements/features of the cloud computing. One example of a data encryption method is *Fully Homomorphic Encryption* (FHE), which is still at the exploring stage.

Other solutions: Other solutions for securing cloud computing include, but are not limited to, *Intrusion Detection System* (IDS), *Cloud Dependability Models* (CDM), *Secure Socket Layer* (SSL), and *Service Level Agreement* (SLA).

Trust Management: Trust management mainly refers to a system that uses social trust entities to ensure secure decision-making and cryptographic credentials. The method is usually associated with access control management.

Students will read more details about security and privacy of mobile cloud computing in Part III.

4.7 SUMMARY

In summary, this chapter talks about the cloud computing architecture as well as its implementations in mobility. The design of the cloud architecture can be based on a variety of aspects. The workload is one of the dimensions of designing cloud architecture. The optimization of hybrid cloud resource management is a crucial part of improving hybrid cloud computing performances. The IWF hybrid cloud management system is a basic optimization approach in hybrid cloud systems. Moreover, security concerns play an important role in cloud architecture designs, too. This is an aspect that is mainly considered by cloud users.

Mobile cloud computing needs to consider not only security issues in non-mobile cloud computing but also threats in mobility. A few dimensions of security issues in clouds include operational abuse on the cloud side, vulnerability caused by non-encryptions, trust issues and multi-tenancy in cloud computing, intrusions in wireless networks, and customer loss of control. We provide an overview of security solutions that will be further discussed in later chapters.

4.8 EXERCISES

1. What is a general goal of a high-performance hybrid cloud resource manager?

2. Briefly describe a four-layer cloud computing architecture.

3. Briefly describe the architecture of mobile cloud computing.

4. When and why do we need the hybrid cloud computing architecture?

5. What is operational abuse on the cloud side considering the security issue in mobile clouds?

6. What are trust and multi-tenancy issues in cloud computing?

7. What issues can be caused by customer loss of control?

8. You have two cloud servers, a remote cloud server and a remote cloud server. The parameters of two cloud servers are shown in Table 4.2. The number of input data is 20. Your mission is to assign 20 data to these two servers and minimize the total cost under a timing constraint 150. Detailed steps are required to explain your solution.

Table 4.2 Data Consumption on Different Cloud Servers for Exercise 8.

	Time Per Data	Cost Per Data
Remote Cloud Server	10	5
Local Cloud Server	5	8

4.9 GLOSSARY

Cloud Resource Manager is a layer in the cloud computing resource management architecture, in which a series of resource operations are accomplished, including resource information retrieval, task assignment, computation process display, communication interconnection, and operation surveillance. This layer can be either physical infrastructure or software-based solutions using virtual operating interfaces.

Function Block is a certain function achieved by a specific cloud resource in distributed cloud computing.

Generic Interface means the manipulative interface operated by the cloud resource manager.

Hybrid Cloud Resource Manager is a typical cloud resource manager that organizes, interconnects, monitors, and facilitates operations for a group of different cloud providers.

Mobile Cloud Computing Architecture is a structure of designing, developing, and implementing mobile cloud computing solutions.

Total Cost of Ownership is a term estimating the direct and indirect consumption of a system investment.

Uniform Resource Identifier is a string of characters identifying the resource name by which the interactions of the resource on the networks are enabled.

II

Optimizations of Data Processing and
Storage in Mobile Clouds

CHAPTER 5

Basic Optimizations: A Perspective of Cloud Computing Performance

CONTENTS

Performance OF CLOUD COMPUTING AND GREEN CLOUDS are two crucial issues in

implementing cloud-based solutions. There are many different parameters that can be considered when cloud designers design a cloud system. Students will understand the crucial aspects of cloud computing performance in the development of cloud systems after reading this chapter. Moreover, students should be able to address the following questions:

1. What does *Performance* mean in cloud computing? Why do we need to consider different dimensions?

2. What does *Green Cloud Computing* mean? What is the basic approach for achieving green clouds?

3. Can you produce a basic algorithm for green clouds?

Focus on the performance of cloud computing in an environmental-friendly way.

5.1 INTRODUCTION

The *Performance* of cloud computing is a status of examining the entire or partial working state of the cloud system, while multiple input factors and outcomes are considered. We use this definition throughout this book. In different application scenarios, cloud computing performance can be considered in various dimensions. A *Dimension* in cloud computing performance, also known as a *Performance Dimension*, refers to a constraint/factor/parameter that needs to be assessed, examined, or considered for the purpose of system optimization. Some common dimensions include service demands, operating environments, infrastructure deployments, and networking conditions. A brief description of each dimension is given as follows:

- *Demands:* A demand is an expected requirement that is needed by either a group of users or the system. A demand dimension usually has a fixed constraint or a range, such as a timing constraint, an energy limitation, or a power level.

- *Operating Environments:* An operating environment, known as the *Integrated Applications Environment* (IAE), may influence the outcomes of cloud computing. Two basic components of an operating environment include a *User Interface* (UI) and an *Application Programming Interface* (API).

- *Infrastructure:* The infrastructure can be also a dimension in the examinations of cloud computing performance. It refers to the amount or capacity of the equipment or facility that supports and delivers cloud services. This dimension needs to be considered when various levels of the infrastructure can result in different outcomes.

- *Networks:* A dimension of a network can refer to a variety of aspects in networking connections. The response time is an important criterion for examining the cloud computing performance from a networking perspective. In addition, the *Network Capacity* is also an important measurement that consists of a group of complex evaluations for knowing the maximum amount of data transmissions under certain networking configurations.

The next section gives a detailed description of cloud computing performance in both performance dimensions and methods.

5.2 CLOUD COMPUTING PERFORMANCE

5.2.1 Two Technical Dimensions

Let's introduce performance-related problems starting with an example in this section. We provide a sample of two dimensions from a technical perspective in order to assist students in understanding performance analysis in cloud computing. Two dimensions in this example include *Response Time* and *Energy Costs*. In practice, there may be other elements acting as dimensions, such as the security level, data volume, and infrastructure alternative.

Figure 5.1 illustrates a scenario in which the cloud resource manager is assigning tasks to the remote cloud servers. The input consists of two tasks, which are *Task 1* and *Task 2*. Two remote cloud servers are available, namely, *Cloud A* and *Cloud B*. The response time and energy costs vary when tasks are assigned to two different cloud servers. The required response time and energy costs are shown in the figure. For instance, according to the tables shown in the figure, *Task 1* requires 5-unit response time on *Cloud A* rather than 7-unit response time on *Cloud B*. From the perspective of performance, we can propose a problem that asks for the lowest energy cost when the timing constraint

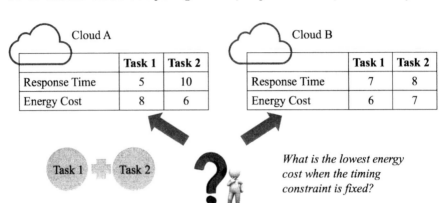

Figure 5.1 Performance problem example: 2-dimensional constraints.

is fixed. In this case, the value of energy cost is the criterion of examining cloud computing performance. In an optimization problem, minimizing the value of energy cost within a given timing constraint is the target of this example. Multiple dimensions can exist in real-world problems. We will introduce a few optimization scheduling algorithms in the following chapters.

5.2.2 Basic Task Scheduling Method

We introduce a basic task scheduling method in this section. In essence, solving a general task scheduling problem needs to follow a series of steps. Figure 5.2 illustrates an example of the basic process that is designed for scheduling tasks to different servers. The server can be either a local or a remote server. The figure also depicts a few basic concepts of task scheduling.

First, there are n input tasks in the figure, *Task 1*, *Task 2*, ..., and *Task n*. A task refers to a requested service form from the clients. Moreover, a *Priority Allocator* (PA) is a node that adjusts or determines the priorities of the input tasks. In practice, some tasks may have a higher-level priority than other tasks; thus, a PA can assist the system in arranging the tasks, according to the task priority levels. Next, the tasks are pushed into a group of queues. A *Queue* is a kind of abstract data type that follows the *First-In-First-Out* (FIFO) data structure. Furthermore, a calculation of costs will be done to determine the individual cost of each queue on every server, which is represented

Input Tasks

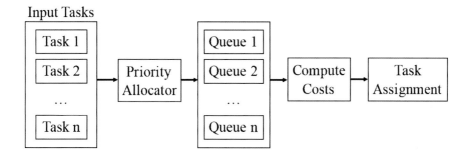

Figure 5.2 An example of the basic process of scheduling tasks to different servers.

as *Compute Costs* in the figure. Finally, a task assignment is generated after a comparison or an adjustment depending on the requirements, such as minimizing the response time or lowering energy costs.

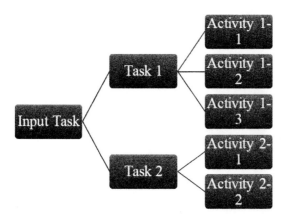

Figure 5.3 Illustration of showing examples of *Tasks* and *Activities*.

Additionally, some complicated input tasks need to be divided into a few small-sized tasks. Figure 5.3 represents an example of dividing tasks into activities. As shown in the figure, the input tasks consist of two tasks, *Task 1* and *Task 2*. Next, *Task 1* is divided into three activities, namely, *Activity 1-1*, *Activity 1-2*, and *Activity 1-3*. The other task, *Task 2* is divided into two activities, which are *Activity 2-1* and *Activity 2-2*. In this example, an activity is configured from dividing the task, so an activity must belong to a task.

5.2.2.1 Use Directed Acyclic Graph

A *Directed Acyclic Graph* (DAG) is a useful tool in solving task assignment problems. In computer science, a DAG is used to model the preceding-succeeding relations depending on the tasks' dependencies, which is represented in topological order. A DAG must be a directed graph, which means that a loop is not allowed in a DAG. Moreover, a DAG is formed by a number vertices and edges, and each edge shows a direction from one vertex to another.

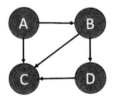

Figure 5.4 An example of a DAG consisting of four tasks.

Figure 5.4 depicts an example of a DAG that has four vertices and five edges. The vertices refer to tasks and the edges refer to the preceding-succeeding relations. For example, from task A to task C, there are three paths, including $A \rightarrow C$, $A \rightarrow B \rightarrow C$, and $A \rightarrow B \rightarrow D \rightarrow C$.

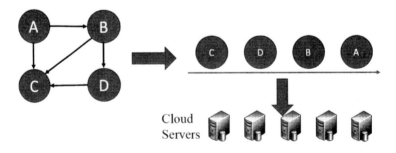

Figure 5.5 Scheduling tasks based on the DAG given in Figure 5.4.

Figure 5.5 represents the process of scheduling tasks based on the DAG shown in Figure 5.4. The sequence {A, B, D, C} derives from the dependencies of all tasks. This procedure can be done by a PA, as mentioned in Section 5.2.2. Next, after a computation, these tasks are assigned to different cloud servers. The following section introduces a fundamental mechanism of modeling cloud performance using a DAG.

5.2.2.2 Basic Cloud Performance Modelization Using the DAG

Section 5.2.2.1 explains the applications of the DAG in cloud computing and a few basic calculations. There are two important aspects addressed in this section. First, we introduce the method of modelizing cloud computing performance by using the DAG. This is a fundamental skill for analyzing cloud computing deployments. Second, we represent a few basic rules when the method of cloud performance modelizations is applied.

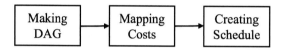

Figure 5.6   Three crucial steps of modelizing cloud performance.

Cloud performance modelizations have three crucial steps, which are displayed in Figure 5.6. The detailed description of the three steps are illustrated below:

1. First, we need to understand the interrelations of the input tasks or activities. This means we need to obtain the dependencies of the activities, by which a DAG is drawn. This procedure is a fundamental operation of formulating the problem in a graphical manner. The dependencies of input tasks/activities need to be accurately shown in the created DAG.

2. Next, we need to calculate the cost for each input task/activity mapped in the DAG. The methods of calculating costs vary and depend on the problems and application scenarios. One of the methods of determining costs is to map each task/activity under all available working modes. Details of mapping costs will be introduced in the following chapters.

3. Finally, a task scheduling plan is created by implementing a resource scheduling algorithm. A *Resource Scheduling Algorithm* is an algorithm that is designed to achieve the expected goals by optimizing the resource usage plans or arrangements. In cloud

computing, a common resource management problem is to minimize the total costs when a group of input tasks are assigned to different cloud servers with various capacities.

In addition, for implementing the method above, we need to follow a few rules:

1. An activity/task can only start after all its immediate predecessors finish. An activity's/task's successor(s) cannot be operated until the activity is finished.

2. We consider the activity/task that does not have immediate predecessor(s) as an *Entry-Activity* or an *Entry-Task* .

3. We consider the activity/task that does not have immediate successor(s) as an *Exit-Activity* or an *Exit-Task*.

The following section provides a few samples of modelizing cloud computing performance evaluations in order to show a good picture of the mechanism.

5.2.3 Examples of Evaluation Modelizations

We represent two simple examples of assessing cloud computing performance in this section. The first example focuses on showing the method of mapping costs, which is covered in Section 5.2.3.1. The other example provides a picture of the main procedures.

5.2.3.1 Mapping Costs

We present a simple example of assessing *Cost Evaluations* (CE) for *Cloud Computing Resources* (R). This is a simple application of obtaining and mapping costs. A complex case can also be solved by using the mechanism used in the given example. For the purpose of the evaluation, a few parameters and variables are involved in this example:

- A refers to the number of total demanded R instances.

- B refers to the number of total initial R instances.

- C refers to the cost evaluation of R that is counted as a percentage.

- D refers to the number of unused R instances.

Our goal is to evaluate the value of C. The formula is $C=A/B$. The value of B derives from $B=A+D$. Figure 5.7 illustrates the paradigm of evaluating costs for cloud computing resources.

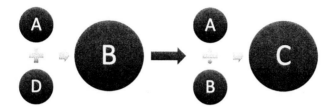

Figure 5.7 Illustration of cost evaluations for cloud computing resources.

5.2.3.2 Calculations of Total Costs

This section presents an example of estimating the execution time. One of the purposes of giving a simple example is to show the basic operations of formulating an execution problem in mathematical expressions. The input task consists of a group of homomorphic activities, denoted as $\{A_1, A_2, \ldots, A_n\}$. Assume that each activity requires the same execution time. The problem is to estimate the execution time of the *ith* activity A_i. The conditions include:

- *T:* The total execution time of the input task.

- *N:* The number of activities in T.

- A_i: The *ith* activity in T.

- *T(i):* The estimated execution time of A_i.

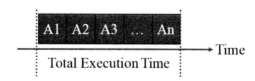

Figure 5.8 Estimated execution time in a timeline.

Figure 5.8 represents the estimated execution time in a timeline. The *Total Execution Time* (TET) sums up the execution time of each activity in the task, represented as $TET = \sum A_i$. A simple division operation can be used, since it is assumed that each activity requires the same execution time period. Therefore, the estimation can be obtained by the following formula: $T(i) = T/N$. The following section presents a case study showing the method of calculating total execution time when heterogeneous cloud computing is applied.

5.2.3.3 Case Study: Calculate Total Execution Time

We have a case study to illustrate a clear picture of assessing cloud computing performance. The conditions of this example include a DAG and a table of cost mapping. The requirement is to calculate the total execution time by using the Greedy algorithm. Figure 5.9 represents the given DAG. As shown in the figure, the input includes four tasks, namely, A, B, C, and D. Task A is the entry-task and D is the exit-task.

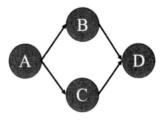

Figure 5.9 The given DAG used in the example (Section 5.2.3.3).

Table 5.1 Cost Mapping of All Tasks on Each Cloud Server. EET: Estimated Execution Time.

	Cloud Servers	A	B	C	D
	Cloud Server 1	10	4	2	9
EET	Cloud Server 2	3	8	9	5
	Cloud Server 3	8	5	4	6

Table 5.1 displays the cost mapping of all tasks on each cloud server. The table also shows that there are three cloud servers available in this example. Tasks have different costs on various cloud servers. For instance, task A needs a 10-unit cost on cloud server 1 but only requires

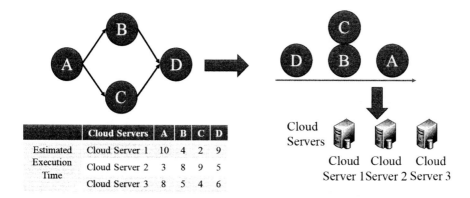

	Cloud Servers	A	B	C	D
Estimated Execution Time	Cloud Server 1	10	4	2	9
	Cloud Server 2	3	8	9	5
	Cloud Server 3	8	5	4	6

Figure 5.10 Paradigm showing the operating sequence based on the given DAG.

a 3-unit cost on cloud server 2. Our objective is using the Greedy algorithm to estimate the total execution time of operating all activities.

The first procedure is to understand the order of task operations. According to Figure 5.9, we know the dependences and relations between activities. A sequence of operating order can be drawn. Figure 5.10 shows the paradigm of the operating sequence, which is based on the given DAG. Task A is the entry-task. Tasks B and C can be executed after task A. Lastly, task D is executed.

Finally, we can produce a task assignment plan by using the Greedy algorithm. Figure 5.11 shows the target plan. We always select the

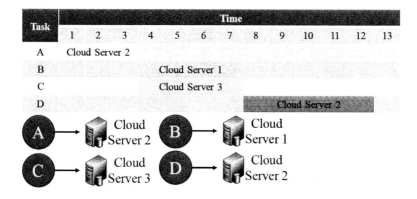

Figure 5.11 Task assignment plan in a timeline table.

shortest execution time for each task on every cloud server, which follows the principle of the Greedy algorithm. The plan is A → cloud server 2, B → cloud server 1, C → cloud server 3, and D → cloud server 2.

An open question is given in Box 5.2.3.3.

Box 5.2.3.3: Open Question

Based on the conditions given in Section 5.2.3.3, can you provide a better solution by using other algorithms? Use the following table for practice.

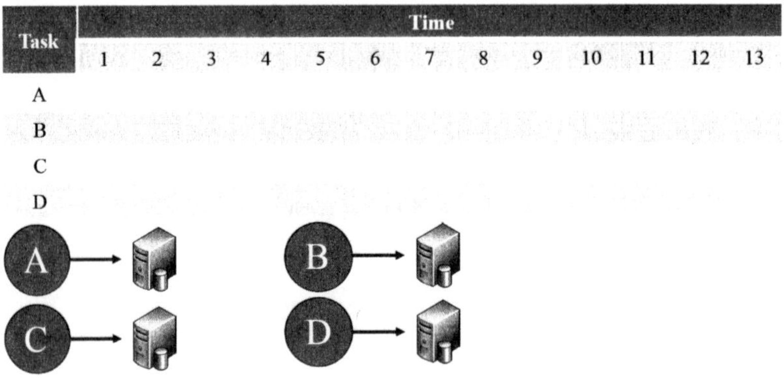

Figure 5.12 Fill up the table by using your algorithm.

The following section introduces green cloud computing, which is another concentration of cloud applications.

5.3 GREEN CLOUD COMPUTING

In this section, we provide a dimension of cloud computing performance optimization, which is green cloud computing. The section is designed to assist students in understanding the mechanism of optimizing cloud-based solution performance for reaching an environmentally friendly target.

5.3.1 Basic Concepts of Green Cloud Computing

Green Cloud Computing (GCC) is a group of cloud-based solutions that are designed to consider both service performance and environmental benefits. Implementing GCC is usually related to defining service models. The major goal of using GCC is optimizing energy efficiency when high performance is achieved. The dimension of high performance needs to be defined by the system designers or users, such as data processing efficiency or data storage capacity.

To achieve the environmental benefits, there exist a few trade-offs between computing resources. A *Trade-Off* in GCC refers to the balance determination of computing resource usage, which can decide the levels of service quality and environmental protections. Some examples of trade-off dimensions include data processing, data storage, or transports. This is the crucial aspect of designing GCC service models, since system developers/designers need to consider the balance between environment elements and performance levels. Figure 5.13 represents a structural figure of GCC with three common dimensions that have strong ties with cloud computing. The common dimensions are business, energy, and technology. The figure also implies two crucial driving forces of applying GCC, which are aligned with business and energy.

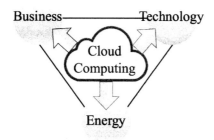

Figure 5.13   Three dimensions of green cloud computing.

First, the energy cost is remarkably increasing, along with the increasing amount of infrastructure and the expansion of applications. Saving energy is a crucial consideration in reducing financial budgets for many cloud practitioners and vendors. This is one of the enablers for enterprises to apply GCC to decrease the energy demands. For reducing the total energy consumptions, a solution is usually associated with resource scheduling, resource allocations, and resource management.

Second, a large volume of energy utilization is usually related to environmental pollution. The critical mission is to reduce the volume of energy consumption. In the cloud system, distributed networks and servers make the energy costs dramatically distributive and complex. For example, the energy consumption caused by the distributed networks would be at a high level, if the data transmission workload is heavy. In this case, an on-premises solution may require less energy. Therefore, a task assignment is required when considering the energy consumption in distributed networks.

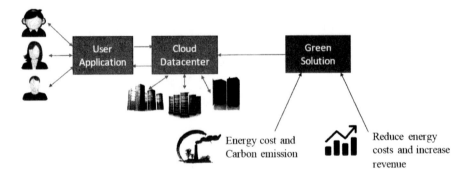

Figure 5.14 A basic green cloud computing model from a datacenter perspective.

Currently, a good example of the GCC implementation is establishing a cloud datacenter. The design focus is the deployment of a cloud datacenter. Figure 5.14 represents a basic GCC model from the perspective of a datacenter. The first method is building up a large datacenter for supporting a great number of users. One technique of developing a large datacenter is applying *Virtual Machine* (VM) consolidations. This technique can enable a great number of cloud users to share one centralized cloud server, such that the number of required cloud servers can be reduced. Some servers can be kept in a sleep mode when the demands go down.

This approach's major benefit is that a resource scheduling algorithm can be applied inside the central server. The main drawback is that the energy cost on networks maybe great due to the lack of controls and messy distributions. In contrast, another method is using smaller distributed data centers that employ servers requiring less energy. This manner is flexible for distributed users. The total energy cost on networks can be reduced, but resource scheduling optimization has rare influences on energy saving.

In summary, a simple goal of using GCC is to produce a cloud computing deployment plan that can achieve the acceptable and applicable service quality, and requires the lowest level of energy supplies. The next section introduces a basic algorithm of green clouds, from algorithm design to assessments.

5.3.2 Dimensions in Algorithms of Green Clouds

In this section, we describe a basic GCC algorithm. We know the defining performance dimension is a fundamental procedure for designing green cloud algorithms from Section 5.3.1. Figure 5.15 shows the architecture of GCC from a datacenter perspective. The figure illustrates four dimension samples determining the resource management in a datacenter. Four dimensions include Internet devices, servers, temperature control facility, and electrical infrastructure. For saving energy, either one or multiple trade-off dimensions can be considered in resource management.

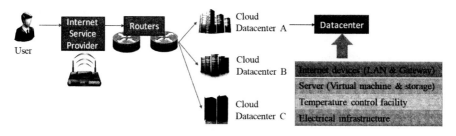

Figure 5.15 Architecture of green cloud computing with four dimensions from a datacenter perspective.

We provide a group of GCC trade-off dimensions considering heterogeneous properties of the system. A brief representation is given as follows:

1. Geographical distributions: Consider the geographical locations; GCC can be impacted by the following elements from this dimension, which include environmental factors, electricity prices, in-

frastructure conditions, and distribution deployments. The trade-off exists when the conditions caused by the geographical distributions are various.

2. Computing configurations: This is a common dimension that is usually associated with distributed cloud servers. The trade-off can exist between different equipment/hardware configurations. For example, *Central Processing Unit* (CPU) configurations can vary, which can result in different computation speed or energy consumption.

3. Energy efficiencies: The heterogeneous properties of energy efficiencies mainly include the available hardware settings that require different levels of resource support or have multiple levels of performance. For instance, distinct wireless networks may require various energy supports.

4. Carbon footprint: The concept of the *Carbon Footprint* (CF) in GCC is the total volume of greenhouse gases that are created by employing or supporting cloud computing activities, services, or operations. The level of CF can be expressed by the equivalent *Carbon Dioxide* (CO_2) counted in tons. This dimension is usually related to the energy supplies to cloud infrastructure, including servers, networks, and user end devices.

The GCC trade-off dimensions summarized above provide a guideline for selecting a design method when formulating energy-saving plans. Furthermore, a few specific dimensions considering energy-aware solutions in GCC have been explored in prior research. We present a few methods of building trade-off models for GCC below.

1. Workload detections: There are two common types of workload detections, which include *Host Overload Detections* (HOD) and *Host Underload Detections* (HUD). HOD is usually supported by a few data analytic techniques, such as deviation and regression [35]. Meanwhile, HUD can be achieved by the technique of VM switches or VM changes, which is considered a resource management problem [36, 25].

2. VM selections: This is a crucial optimization issue in achieving high performance when green clouds are considered. The purpose of optimizing VM selections is to complete the input jobs within

the configured constraints. For instance, a time-aware algorithm designed for GCC achieves an environmentally friendly solution that can be completed within a timing constraint. The Greedy algorithm and *Random Selection Policy* (RSP) are two general options for generating a strategy or plan of selecting VMs. An example of RSP is *First-In-First-Out* (FIFO). Finally, for those complex input tasks with complicated parameter settings, a *Maximum Correlation Policy* (MCP) is suggested. Based on the conditions and requirements, we can apply different resource management algorithms to solve these problems in practice, such as Greedy and dynamic programming [4, 27, 25, 37].

3. VM placement: A *Virtual Machine Placement* (VMP) refers to the selection process of determining the host of VMs. The selection of VM usually matches the placement goals made by developers/users. The results derive from a few parameters or variables, such as forecasting usage, required resources, available infrastructure, and hardware capacities. One approach of implementing VMP is rating hosts. The administrator chooses the VM's host based on the rates of all available hosts.

Understanding the performance dimensions is a fundamental step in creating an algorithm for GCC. System designers need to know where the balance points are when different factors are considered. Section 5.3.3 presents a simple method of creating GCC algorithms, as well as a motivational example.

5.3.3 Creating an Algorithm for Green Clouds

5.3.3.1 Crucial Steps

In this section, we introduce a basic method of creating GCC algorithms, which consists of three crucial procedures. Figure 5.16 illustrates a flow diagram showing the basic steps of designing the energy-aware scheduling algorithm. Three main steps include developing a green cloud manager, analyzing requirements, and scheduling tasks, as shown in the figure. Each step has a few sub-tasks.

At the step of the requirement analysis, listing cloud parameters is a critical task by which the trade-off dimensions are determined. First, the system designers need to accomplish the parameter/variable collections, such as carbon emission rate and CPU power efficiency.

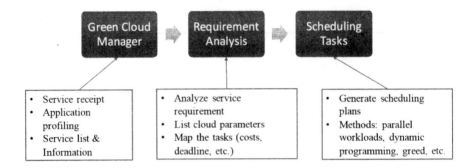

Figure 5.16 A diagram of the main steps for designing the energy-aware scheduling algorithm.

Based on the expected service demands, the next mission is to complete a group operation to sort the tasks. The sorting operation is an implementation of the designed algorithms by utilizing the obtained variables/parameters, such as deadlines and CF. We use CE to denote *Carbon Emission*, CDE to denote *Cloud Datacenter Efficiency*, and VMEE to denote *VM's Energy Efficiency*. The method of calculating CF is given by the following Eq. (5.1).

$$CF = CE \times CDE \times VMEE \tag{5.1}$$

The following section shows a simple example of using an energy-aware scheduling algorithm that uses the equation above.

5.3.3.2 Sample Energy-Aware Scheduling Algorithm

In this section, we provide an example of showing the difference between the energy-aware scheduling algorithm and the normal Greedy algorithm. The conditions of this example include a group of input tasks, as well as their costs on different cloud servers. There are four input tasks in this example, which are $T1$, $T2$, $T3$, and $T4$. The DAG is shown in Figure 5.17.

Moreover, there are three cloud servers available, $C - A$, $C - B$, and $C - C$. The required execution time for each task is mapped in Table 5.2. The requirement for the first task assignment plan is producing a performance-focused solution, which means all tasks need to be finished as early as possible. Thus, the execution time is mainly considered during plan generation. The requirement for the second task

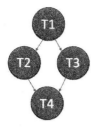

Figure 5.17 DAG used by the sample in Section 5.3.3.2.

assignment plan is to produce a green-focused solution, which needs to use CF to determine the plan. We will use the Greedy algorithm to produce both plans in order to compare the results.

Table 5.2 The Required Execution Time for Four Input Tasks on Three Clouds.

	C-A	C-B	C-C
$T1$	5	6	7
$T2$	4	8	5
$T3$	2	5	6
$T4$	2	3	3

First, we use the Greedy algorithm to complete the performance-focused task scheduling plan. According to the operating principle of the Greedy algorithm, we produce the task assignment plan shown in Table 5.3. This approach does not consider the influence of CF. The total execution time is 16, derives from (5+4+5+2).

Table 5.3 Task Assignment Plan by Using the Greedy Algorithm for the Example in Section 5.3.3.2.

T1	→	C-A
T2	→	C-A
T3	→	C-B
T4	→	C-A

Next, we develop a green-focused Greedy algorithm by considering the impacts of CF. The first step is obtaining the parameter values, such as CE, CDE, and VMEE, as displayed in Eq. (5.1). Table 5.4 shows a mapping that provides all values of the required parameters. The tasks have different parameter values on various clouds. In general, the information shown in Table 5.4 is the condition of producing the task assignment. The methods of gaining the information vary, such as developing a prediction mechanism or classifying tasks into layer-based groups.

Table 5.4 Parameter Table for CE, CDE, and VMEE.

Task	Cloud	CE	CDE	VMEE
	A	7	0.8	0.5
T1	B	5	0.6	0.4
	C	6	0.5	0.9
	A	8	0.4	0.6
T2	B	5	0.8	0.4
	C	6	0.4	0.7
	A	4	0.8	0.8
T3	B	5	0.6	0.7
	C	5	0.9	0.4
	A	8	0.6	0.5
T4	B	7	0.5	0.7
	C	9	0.8	0.7

In addition, we need obtain the CF value of each task on each cloud server. Table 5.5 shows the calculation operations deriving from Table 5.4. The partial Table 5.5 located to the right of the right arrow lists the calculation results.

Furthermore, we can produce another task assignment plan by using the results given in Table 5.5. We consider the value of CF as the priority factor, such that the results are different from the previous one. The generated plan is shown in Table 5.6. The execution time of this approach is 22, derives from (6+8+6+2), which is longer than the performance-focused Greedy, 16. Meanwhile, the green-focused solution for total CF value is 7, which is much less than the result gained by the performance-focused solution, which is 9.22.

Table 5.5 Mapping CF for Green Cloud Computing.

Task	Cloud	CE	CDE	VMEE		Task	Cloud	CF
	A	7	0.8	0.5			A	2.8
T1	B	5	0.6	0.4		T1	B	1.2
	C	6	0.5	0.9			C	2.7
	A	8	0.4	0.6			A	1.92
T2	B	5	0.8	0.4		T2	B	1.6
	C	6	0.4	0.7	$\Rightarrow$		C	1.68
	A	4	0.8	0.8			A	2.56
T3	B	5	0.6	0.7		T3	B	2.1
	C	5	0.9	0.4			C	1.8
	A	8	0.6	0.5			A	2.4
T4	B	7	0.5	0.7		T4	B	2.45
	C	9	0.8	0.7			C	5.04

Table 5.6 Task Assignment Plan Generated by Using the Greedy Algorithm when GCC is Considered for the Example in Section 5.3.3.2.

T1	$\rightarrow$	C-B
T2	$\rightarrow$	C-B
T3	$\rightarrow$	C-C
T4	$\rightarrow$	C-A

In summary, we provide an example showing the features of the GCC-based approach in this section. According to the illustration of the example, we understand that the performance of cloud systems may be lowered if the environmental elements are considered. This is also an example of showing trade-off dimensions. Moreover, in some other situations, the system is required to reach a performance requirement, as well as use a constrained volume of energy. Using Greedy algorithms can efficiently generate a plan; however, the outcome is usually not an optimal solution [25, 38]. For increasing the entire system's performance, some other algorithms can be applied, such as the genetic algorithm [6, 7, 39]. For obtaining an optimal solution, system designers may use dynamic programming to design an algorithm [40, 4, 41, 42].

5.4 FURTHER READING

1. M. Qiu, Z. Ming, J. Li, K. Gai, and Z. Zong. Phase-change memory optimization for green cloud with genetic algorithm, *IEEE Transactions on Computers*, vol. 64, no. 12, pp. 3528–3540, IEEE, 2015.

2. K. Gai, Z. Du, M. Qiu, and H. Zhao. Efficiency-aware workload optimizations of heterogeneous cloud computing for capacity planning in financial industry. In *Proceedings of The IEEE 2nd International Conference on Cyber Security and Cloud Computing*, pp. 1–6, New York, USA, IEEE, 2015.

3. K. Gai, M. Qiu, H. Zhao, L. Tao, and Z. Zong. Dynamic energy-aware cloudlet-based mobile cloud computing model for green computing, *Journal of Network and Computer Applications, 59*, vol. 59, pp. 46–54, Elsevier, 2016.

4. K. Gai, M. Qiu, H. Zhao and M. Liu. Energy-aware optimal task assignment for mobile heterogeneous embedded systems in cloud computing, in *Proceedings of The 2016 IEEE 3rd International Conference on Cyber Security and Cloud Computing (CSCloud)*, pp. 198–203, Beijing, China, IEEE, 2016.

5. K. Gai and M. Qiu and X. Sun and H. Zhao. Smart data deduplication for telehealth systems in heterogeneous cloud computing, *Journal of Communications and Information Networks*, vol. 1, no. 4, pp. 93–104, Springer, 2016.

5.5 SUMMARY

In this chapter, we mainly talk about cloud computing performance and its relevant aspects in GCC. A few basic methods of scheduling tasks are introduced in this chapter, from using DAG to modelizing cloud performance, from mapping costs to calculating the total execution time. Moreover, GCC is an important aspect of evaluating the performance of adopting cloud computing. We emphasize the significance of understanding trade-off dimensions in this chapter. A number of relevant concepts are described in this chapter, too. Students should have a good understanding of cloud computing performance and the concept of trade-off dimensions, which is an important fundamental for designing advanced cloud systems.

5.6 EXERCISES

1. What does *Performance* mean in cloud computing?

2. What is a dimension in cloud computing performance? List two common dimensions in examining cloud computing performance.

3. Discuss an optimization problem when considering both *Response Time* and *Energy Costs*.

4. What is a *Priority Allocator* (PA) in task scheduling methods?

5. Briefly describe the basic process of scheduling tasks to different cloud servers.

6. What is a *Directed Acyclic Graph* (DAG) and how can it be used in cloud computing optimization problems? Describe a basic mechanism of using a DAG to model the cloud performance optimization problem.

7. What is the concept of *Green Cloud Computing* (GCC)? Which dimensions are considered in this concept? What are trade-offs?

8. Exercise 8:

   You have 4 input tasks and 3 cloud servers. The required execution time of 4 input tasks on 3 cloud servers is given in Table 5.7. The DAG for 4 input tasks is shown in Figure 5.18. All parameters regarding CE, CDE, and VMEE are given in Table 5.8. Your mission:

   (a) Use the Greedy algorithm to develop an efficiency-oriented scheduling algorithm. The optimization algorithm needs to obtain a solution to reduce the total execution time.

Table 5.7 The Required Execution Time of 4 Input Tasks on 3 Clouds for Exercise 8.

	C-A	C-B	C-C
T1	4	7	9
T2	3	6	4
T3	3	8	7
T4	3	5	6

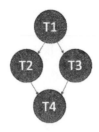

Figure 5.18 DAG for Exercise 8.

Table 5.8 Parameter Table of CE, CDE, and VMEE for Exercise 8.

Task	Cloud	CE	CDE	VMEE
	A	8	0.7	0.6
T1	B	6	0.7	0.5
	C	3	0.4	0.9
	A	7	0.6	0.8
T2	B	4	0.8	0.5
	C	7	0.5	0.5
	A	6	0.7	0.8
T3	B	4	0.7	0.5
	C	4	0.6	0.7
	A	9	0.7	0.5
T4	B	7	0.3	0.5
	C	8	0.6	0.8

(b) Use the Greedy algorithm to develop an energy-aware scheduling algorithm. The optimization algorithm needs to reduce the total carbon footprint.

(c) Compare the results gained from the two algorithms and discuss.

9. Exercise 9:

In this exercise, a DAG is given ins Figure 5.19. The tasks' costs on each cloud server are displayed in Table 5.9. The following questions need to be answered:

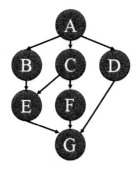

Figure 5.19 DAG for Exercise 9.

Table 5.9 Cost Mapping of All Tasks on Each Cloud Server for Exercise 9*.

	CS	A	B	C	D	E	F	G
	CS1	3	5	6	2	4	6	9
EET	CS2	6	2	4	7	6	5	3
	CS3	4	4	9	7	8	5	4

* EET: Estimated Execution Time. CS: Cloud Server.

(a) How long is the execution time?

(b) What is the task assignment by using your algorithm?

(c) Is your algorithm an optimal algorithm? If yes, provide the proof.

5.7 GLOSSARY

Carbon Footprint (in GCC) is the total volume of greenhouse gases that are created by employing or supporting cloud computing activities, services, or operations.

Demand (dimension) is an expected requirement of a group of users or the system.

Dimension refers to a constraint/factor/parameter that needs to be assessed, examined, or considered for the purpose of system optimizations (in cloud computing performance).

Entry-Activity/Task is the activity/task that does not have immediate predecessor(s).

Exit-Activity/Task is the activity/task that does not have immediate successor(s).

Green Cloud Computing is a group of cloud-based solutions that are designed to consider both service performance and environmental benefits.

Network Capacity is a measurement dimension that consists of a group of complex evaluations for knowing the maximum amount of data transmissions under certain networking configurations.

Performance (in cloud computing) is the status of the entire or partial working state of the cloud system when multiple input factors and outcomes are considered.

Trade-Off (in GCC) refers to the balance determination of computing resource usage, which can decide the levels of service quality and environmental protections.

Virtual Machine Placement refers to the selection process of determining the host of VMs.

CHAPTER 6

Preemptable Algorithm Execution in Mobile Cloud Systems

CONTENTS

Performance OF CLOUD SERVICES is a critical issue in optimizations. In this chapter, we will introduce an important optimization algorithm in cloud systems, which is the preemptable algorithm. The algorithm is designed to optimize the cloud resource allocations. Students will able able to answer the following questions after reading this chapter.

1. What is a preemptable scheduling problem in IaaS?

2. What are the main steps of the cloud resource allocation mechanism?

3. What are advance reservation and best-effort tasks?

4. How can we design a preemptable algorithm to solve a cloud computing optimization problem?

6.1 INTRODUCTION

As we discussed, cloud computing can be considered a single or a few clusters of distributed computers that provide on-demand computational resource service offerings. In current implementations, *Infrastructure-as-a-Service* (IaaS) is usually delivered by leasing computing resources to cloud users. The computing infrastructure is provided to customers using virtualization techniques that facilitate flexible and efficient services. For instance, there are a variety of cloud service leases available at Amazon's Elastic Compute Cloud. Cloud users can customize the scope of the infrastructure services. One setting is that each *Virtual Machine* (VM) provisions the data processing service by the following computing configurations, including a 1.2-Ghz

Opteron processor, 1.7 GB of memory, and 160 GB of disk space. Customers can determine how many VMs they need to lease based on floating practical demands.

This service type can help cloud clients gain "unlimited" computing infrastructure from cloud providers. However, the reality is that the capabilities of the data centers always have limitations, from a technical perspective. The operations on the cloud side are masked by VMs. Overflowing some computation workloads to other data centers is an option for IaaS vendors to extend or scope up the infrastructure resource pools and capabilities. The overflow operations can be done between private and public clouds or within a hybrid cloud. The basic mechanism of overflowing workloads is forwarding some tasks to other data centers when the operations can increase the computation efficiency or costs, which depends on the optimization dimensions.

Despite of a few advantages of forwarding workloads to other data centers, there is a crucial challenge for executing this mechanism. Using the traditional method will encounter challenges caused by the restrictions of resource monitoring and management in that the traditional methods are mainly designed for supporting an operating environment relatively unified to computing resources. Sometimes, using heterogeneous cloud computing is also an alternative for minimizing costs. A flexible resource increase or decrease can lower the execution cost by diminishing the idle cloud infrastructure.

However, forwarding tasks to another data center will face the problems caused by heterogeneous cloud computing, because the traditional method may not be able to work effectively in a hybrid cloud environment. The management mechanism of resource allocations is a crucial part of this issue. This chapter will introduce an advanced solution to this problem, which uses *Preemptable Scheduling* algorithms to solve the problems of forwarding tasks in the heterogeneous cloud context.

6.2 PREEMPTABLE SCHEDULING IN IAAS

The concept of *Preemptable Tasks* in cloud computing refers to an ongoing task that can be temporarily suspended when another input task having a higher-level priority is inserted into the task sequence and the task needs to be resumed after the inserted task finishes [25, 43, 24, 27]. This section introduces implementations of preemptable tasks for overflow tasks in cloud computing.

6.2.1 Basic Cloud Resource Allocation Mechanism

In this chapter, we are going to discuss the basic cloud resource allocation mechanism. A fundamental structure of the mechanism is described to help students to be aware of the operating principle of resource allocations in cloud computing. Moreover, the main steps of the resource mechanism in cloud computing will be summarized.

6.2.1.1 Fundamental Structure of Cloud Resource Allocation Mechanism

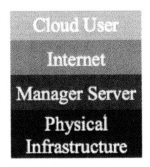

Figure 6.1 Four basic layers of cloud resource allocations.

The fundamental structure of the cloud resource allocation elementarily consists of four layers, including cloud users, the Internet, the manager server, and the physical infrastructure provided by cloud vendors. Figure 6.1 represents a layer-based structure graph illustrating four basic layers. Among these four layers, the *Physical Infrastructure* layer is the objective for executing applications, to which the tasks are assigned. The manager server will abstract the capacity characteristics as well as performance parameters from each cloud. The diversity of server performance makes the whole cloud system a heterogeneous system. Moreover, for the purpose of efficient calculations, we usually consider each cloud server a homogeneous system, even though the computing nodes may be varied due to different hardware deployment.

Figure 6.2 is an example of the fundamental structure for the cloud resource allocation mechanism. The figure shows that there are a few cloud server providers whose service offerings are varied. Multiple clouds provide heterogeneous VMs. The providers are interconnected via the manager server on the Internet, through which cloud users access services.

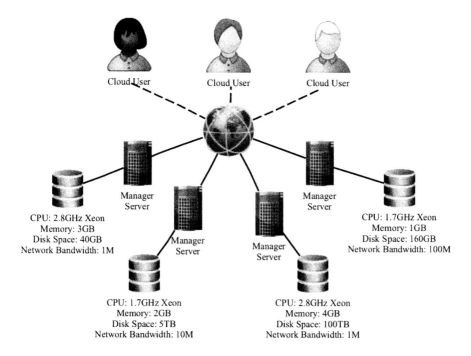

Figure 6.2 An example of the fundamental structure for cloud resource allocation mechanism.

In this example, the displayed information about four servers represents the performance differences in four dimensions, including *Central Processing Unit* (CPU), memory, disk space, and network bandwidth. These different server offerings may require various leasing prices. Therefore, reducing the total execution cost is possible for both cloud users and vendors. One essential approach is dividing the tasks into a group of linked tasks and assigning each task to the server requiring a lower-level cost. The deployment of the manager server provisions the infrastructure coordinations for the purpose of optimization.

6.2.1.2 Main Steps of Cloud Resource Allocation Mechanism

First, one critical role of cloud resource allocation is the *Cloud Manager Server*. This is a cloud resource management entity in cloud computing, by which a series of resource operations are accomplished, including resource information retrievals, task assignments, computation process displays, communication interconnections, and operation surveillances.

This layer can be either physical infrastructure or software-based solutions using virtual operating interfaces.

Moreover, the crucial target of the cloud resource allocation mechanism is to minimize the cost, which needs the manager server to cover a variety of aspects. The target of the manager server is threefold: 1) allocating computing resources to tasks, 2) identifying the order of the task executions, and 3) scheduling overheads from preparations, switching tasks, to terminating tasks. There are mainly four steps for the cloud resource allocation mechanism in scheduling preemptable tasks. The four steps include *Partition*, *Schedule*, *Assignment*, and *Execution*. The detailed statements are given as follows:

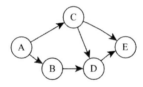

Figure 6.3 An example of the *Directed Acyclic Graph* (DAG) after partitioning. Five tasks include A, B, C, D, and E. Entry task: A. Exit task: E.

1. Partition: This step divides the inputs into a few tasks by using the DAG method, such that the sub-tasks can be considered a set of discrete linked processes. This method can identify the tasks in order of precedence. Figure 6.3 represents an example of the DAG after partitioning. There are five tasks in the figure, including A, B, C, D, and E. The entry task is A and the exit task is E.

2. Schedule: The partitioned tasks need to be scheduled by considering condition parameters, such as timing constraints and preemptable tasks. At this step, the cloud manager server needs to be aware of the cloud resource availabilities in the system and determines whether the available cloud meets the task's requirement. If a cloud is selected, the tasks will be assigned to this cloud and enter a queue in which they await execution. A schedule is a plan for assigning tasks to the best-alternative cloud server.

3. Assignment: The cloud manager server assigns tasks followed by the completed schedule. The tasks can be assigned either to other

clouds or to its own cloud in which the tasks will be stored in a queue. Figure 6.4 illustrates an example of the fundamental structure of the cloud resource allocation mechanism, which is associated with the DAG in Figure 6.3. In this example, the applications are partitioned into five tasks, A, B, C, D, and E. The precedence order of these five tasks is shown in Figure 6.3. Tasks A, C, and E are assigned to the same cloud server. Meanwhile, tasks B and D are assigned to two separate cloud servers.

4. Execution: The tasks are executed on the cloud side physical machine.

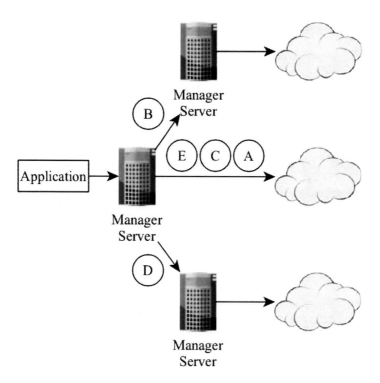

Figure 6.4 An example of the fundamental structure of the cloud resource allocation mechanism.

Among these four steps, the main crucial step for optimizations is step 3, *Schedule*. The improvement of the cloud systems will depend on whether the schedule can reach the desired goal, such as reducing execution time or saving energy consumption. The steps *Assignment* and *Execution* will follow the outcomes of the *Schedule*. The remaining sections of this chapter will introduce the method of scheduling preemptable tasks.

6.2.2 Messaging Methods: Pull-Push Modes

For the purpose of resource scheduling, resource monitoring between cloud customers and service producers is necessary. Operating this mechanism is required by the distributed tasks in cloud computing in order to obtain the availability information of the computing resources. This is because there is no centralized super node deployed in the cloud system; thus the centralization-oriented infrastructure monitoring method cannot satisfy the messaging demands in a distributed system. Two typical messaging methods in resource monitoring include the *Pull* and *Push* modes.

- **Pull Mode:** The customers gain the information about the computing resource status from the producers in a *Pull* messaging mode. The main advantage of using Pull mode is that the transmission cost can be controlled at a low level if the interval of the inquiries is configured at a proper scope.

- **Push Mode:** The producers send/deliver the computing resource status information to the customers when the resource status is updated in a *Push* messaging mode. The virtue of using Push mode is to gain higher-level accuracy as the status update threshold is well defined.

In preemptable tasks scheduling for the resource allocations, both Push and Pull modes are combined for the purpose of efficient communications. The key operation of using Push and Pull modes in resource allocation is being aware of the customers and producers. The *Customer* is the cloud manager server that assigns an application/task to another cloud manager server. Meanwhile, the *Producer* is the cloud manager server that receives the assignment of the application/task from another cloud manager server, which is known as a *Customer*. Figure 6.5 represents a fundamental relationship between the customer and the producer. The manager server of Cloud A is the customer in the figure. The server receiving the application assignment is the producer, which is the manager server of cloud B.

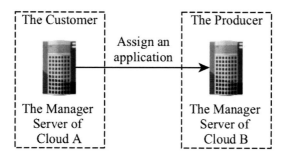

Figure 6.5 The relationship between the *Customer* and the *Producer* in the resource allocation mechanism.

The other important aspect about the customer and producer is the occasion of producer status updates. There are two essential occasions when the customer needs to know the producer's status:

1. The customer needs to know the producers' resource statuses when the customer is about to assign a task/application to other clouds (producers). In Figure 6.5, Cloud A's manager server needs to know the status of Cloud B.

2. The producer needs to inform the customer of the accomplishment of the assigned task/application. As shown in Figure 6.5, Cloud A needs to be informed by Cloud B when the task is finished at Cloud B.

Considering the two occasions above, the executions of the *Pull-Push Mode* are summarized as follows:

- *Pull execution:* The customer pulls the status information of other cloud resources to check the availability of the computing resources. The scheduling allocations will depend on the available computing resources. In reality, the producer may not be able to share their server status information in detail due to security or performance concerns. The operation of Pull needs to be efficient and cooperation-oriented, which means complex operations should be avoided in Pull executions. One method is to pull the earliest available time of each cloud server without any reservation or occupation operations.

- *Non-Pull execution:* The customer does not pull the information concerning the task/application after this task/application is already assigned to a cloud.

- *Push execution:* The producer will push the information to the customer after the assigned task/application is accomplished. The producer will not push any information to the customer before the assigned task/application is finished.

6.2.3 Concepts of the Resource Allocation Model in Cloud Computing

6.2.3.1 Advance Reservation vs. Best-Effort Tasks

Considering preemptable tasks scheduling, there are two typical modes for leasing cloud infrastructure in cloud computing, which include *Advance Reservation* (AR) and *Best-Effort* (BE). In the resource allocation model, some applications use the AR mode, and the other applications are executed in the BE mode. These two task modes are preconditions of implementing preemptable tasks.

- *Advance Reservation* (AR): The tasks have priorities for execution by reserving resources in advance. AR tasks usually have a specific start/available time that is known as the arrival time. The AR task needs to wait in a waiting queue if the assigned cloud server has already been assigned an AR task that is being executed. For the requirement of resource allocations, AR mode also requires the execution time.

- *Best-Effort* (BE): BE tasks are those tasks that follow the *As-Soon-As-Possible* principle for the queue placement. The priority level of BE tasks is lower than AR tasks. Similar to AR tasks, BE tasks need to estimate the time length of the execution.

It is important for us to understand the meaning of both AR and BE tasks. The key points of AR tasks are that they have a higher priority of execution than BE tasks and AR tasks usually have an expected arrival time that is either a zero or a time value. A zero means that the AR task is assigned immediately after the task is input. A time value means that the AR task will be assigned by the manager server after an estimated time period. The arrival time acquisition can be determined by estimations, configurations, or schedules.

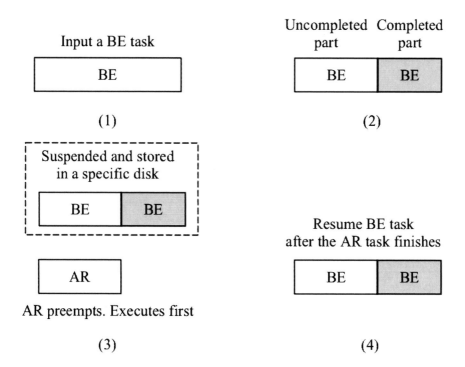

Figure 6.6 The preempting procedure when the AR task interrupts the BE task.

Meanwhile, the key point of the BE task is that the task will be suspended when it is interrupted by an AR task. When the suspension occurs, the VM will suspend the BE task and store the disk image of the BE task in a specific disk space. This process happens before the interrupted AR task preempts the VM. The BE task will be resumed after the AR tasks completes. Figure 6.6 represents the preempting procedure when the AR task interrupts the BE task.

Figure 6.6 (1) shows that there is an input BE task starting the execution. Figure 6.6 (2) shows that the BE task execution is in progress. The gray part is the completed component of the task. Next, Figure 6.6 (3) illustrates that the BE tasks is suspended by being copied and stored in a specific disk while an AR task preempts. The disk image of the AR task is transferred to the VM after the BE task is suspended. Finally, Figure 6.6 (4) shows that the BE task is resumed to the executions by copying from the specific disk after the AR task finishes. The rest of the BE task will be executed.

6.2.3.2 Formulations of the Resource Allocation Problems

Now we understand the main task modes in resource allocations via VMs. The next important knowledge that we need for resource allocations is formulating the problem. This section gives a couple of examples showing the method of formulating a problem in various ways. In general, the formulation derives from the feature extractions that can represent the crucial characteristics of the system. For IaaS cloud systems, we can use a tuple to represent a leased allocated resource in cloud computing. For example, a tuple $\langle n, m, d, b \rangle$ can be used to formulate the leased data processing resources. The meanings of n, m, d, and b are given as follows:

- $n:$ the number of CPUs

- $m:$ memory in megabytes

- $d:$ disk space in megabytes

- $b:$ network bandwidth in megabytes/sec

One assumption that is often made for simplifying the problem formulation is considering that a task requires the same leased allocated resource $\langle n, m, d, b \rangle$ no matter on what type of VM the task runs.

Furthermore, cloud users acquire cloud services via VMs in which the actual computing operations are hidden. In IaaS, one of the methods of counting services depends on the usage of the VMs. When we look at the scene behind the VMs, the service measurement methods used by cloud manager servers are different. A number of tasks can be executed by the same server since a server is able to synchronously support a number of VMs. The cloud manager server can either assign the task to its cloud server or forward the task to other cloud servers. Whether or not the task is assigned to its own cloud server depends on two aspects, which are the server's performance and capacities.

The diverse performance will determine whether the cloud manager server forwards the task to another cloud server. Forwarding a task to other cloud servers usually aims to save costs or increase working efficiency. The server capacity is the other important parameter for determining whether the task should be forwarded. The task can be added to the waiting queue only when the server has sufficient capacity. For example, assume that there exists an input application A consisting of j tasks T_i, $i \in [1, j]$. The tasks need to be scheduled to a server S.

The number of VMs required for each task T_i is $V(T_i)$. The available server capacity (or remaining workload capacity) of the server S is $R(S)$. In this case, the condition of executing all tasks T_i at server S is the value of $R(S)$ that is no smaller than the sum of all required VMs. The formulation is given in Equation (6.1).

$$R(S) \geqslant \sum_{i=1}^{j} V(T_i) \tag{6.1}$$

This is also a good example of formulating the problem. We can articulate a research problem by using mathematical expressions if the parameters and variables are appropriately selected and defined.

In addition, another useful technique in formulating/solving resource allocation problems is applying *Directed Acyclic Graphs* (DAGs) to represent and analyze applications. For example, a DAG $T = (V, E)$ has a number of vertices V. Each V represents a dependent task. An edge E represents the precedence relations between tasks. The task that has no preceding task (predecessor) is an *Entry Task*. The task that has no succeeding task (successor) is an *Exit task*. Figure 6.7 is an example of using a DAG for displaying the task procedure. A node is an entry task and D is an exit task. Task B is called an immediate successor of Task A. And Task A is called the immediate predecessor of Tasks B and C.

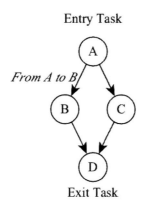

Figure 6.7 An example of a DAG. Entry Task: A; Exit Task: D.

6.2.3.3 Parameters/Variables of the Execution Time

The next technique that is often used for characterizing computational capacities is the *Execution Performance Matrix* (EPM). An EPM in cloud computing is applied for identifying the differences of the execution time between VMs. For instance, we can use a matrix $|M|=|T \times V|$ to evaluate the execution time matrix. T refers to the types of tasks and there exists $T = \{t_1, t_2, \ldots, t_i\}$. V refers to the types of VMs and there exists $V = \{v_1, v_2, \ldots, v_j\}$. The matrix $|M|$ is shown in Equation (6.2). In the matrix, $ET(t, v)$ represents the execution time length for each task on a VM. For example, the execution time of task t_4 on VM v_3 is $ET(t_4, v_3)$. In practice, multiple expressions can be applied, such as a table or mathematical expressions.

$$
M = \begin{matrix} & t_1 & t_2 & \ldots & t_i \\ v_1 \\ v_2 \\ \ldots \\ v_j \end{matrix}
\begin{pmatrix}
ET(t_1, v_1) & ET(t_2, v_1) & \ldots & ET(t_i, v_1) \\
ET(t_1, v_2) & ET(t_2, v_2) & \ldots & ET(t_i, v_2) \\
\ldots & \ldots & \ldots & \ldots \\
ET(t_1, v_j) & ET(t_2, v_j) & \ldots & ET(t_i, v_j)
\end{pmatrix}
\tag{6.2}
$$

Moreover, in resource scheduling mechanisms, the length of the execution time is an important variable that is usually used for evaluating the cloud system's performance. We will introduce a few common execution time variables in this section.

- *Average Execution Time* (AET): This variable is used to represent the average execution time of a certain task t_i. We can express the task t_i's average execution time as $AET(t_i)$.

- *Earliest Start Time* (EST): This variable represents the estimated earliest time for executing the task. The estimated time is based on the predecessor tasks' accomplishment time, which means the task can only start after all its predecessors finish. The mathematical expression is given in Equation (6.3).

$$
EST(t_i) = max\{EST(t_j) + AET(t_j)\}
$$
$$
\text{where } t_j \in \{\text{Predecessors}(t_i)\} \tag{6.3}
$$

In the equation, *max* refers to selecting the maximum value from the target set. Task t_j is a task belonging to the set of t_i's predecessors.

- *Latest Start Time* (LST): This variable refers to the estimated latest time for executing the task. This estimated time derives from the latest execution start time of the successors and the task's average execution time. The value of the task t_i's LST can be acquired by counting the difference between the successor's latest start time and t_i's average execution time. The minimum value will be selected because the successors can be executed only when t_i finishes. The mathematical expression of LST acquisition is given by Equation (6.4):

$$LST(t_i) = min\{LST(t_j) - AET(t_i)\}$$
$$\text{where } t_j \in \{\text{Successors}(t_i)\} \tag{6.4}$$

We can also arrive at a conclusion from this definition, which is that the exit task's LST is its EST since the exit task does not have a successor.

- *Critical Node* (CN): This term refers to a set of tasks in which the values of EST and LST are the same.

6.2.4 Summary

In summary, this section introduces the basic concepts used in the cloud resource allocation mechanisms. The operating principle of the IaaS optimizations is explained. The following sections will describe the three main algorithms for optimizing resource allocations in cloud computing.

6.3 RESOURCE ALLOCATION ALGORITHMS IN CLOUD COMPUTING

In this section, we give three algorithms that can be implemented in resource allocations of cloud computing. The implementations of the algorithms use the techniques introduced in Section 6.2.

6.3.1 Round-Robin (RR) Algorithm

6.3.1.1 RR Algorithm Description

The *Round-Robin* (RR) algorithm is one of the simplest scheduling algorithms, and considers each allocation objective equally and assigns the tasks in the circular order [44]. Executing the RR algorithm does not consider the levels of priority such that this approach follows the *First-Come-First-Serve* rule. In the cloud resource allocation mechanism, the RR algorithm can also be used for solving the problem of scheduling preemptable tasks. The main input of the RR algorithm is a list schedule P that uses the priority-based method based on the given DAG(s). We provide one of the approaches of generating a P and the main steps are given as follows:

1. Initialize a task list P (an empty list) and an empty stack S.

2. Calculate the values of EST and LST for each task.

3. Push the CN tasks into the stack S using the decreasing sequence of LST.

4. The following operations will be executed **While** the stack S is not empty. (Enter a While loop).

    (a) Assign the immediate predecessors to the stack S if S's top has un-stacked immediate predecessors.

    (b) For other tasks, assign the top task of the stack S to the list P by popping out the top task of S.

5. Repeat the process till the stack S is empty. Output the list P when the **While** loop ends.

The method of producing the list schedule can be also used for other scheduling algorithms in this chapter. For the RR algorithm, the list schedule is the input. The output of the RR algorithm is a static resource allocation plan. A brief description of the RR algorithm is given as follows:

Require: a list schedule P using priority-based method from DAG(s)

Ensure: a static resource allocation plan.

1. The cloud manager server searches all available cloud servers and marks them in a sequence.

2. The input tasks will be assigned by the manager server to the cloud server according to the order of the sequence.

3. When one application's tasks are finished, the order of the cloud servers will be started over.

This algorithm is easy to implement, and can be applied in many scheduling problems. It is also easy for cloud developers to establish the network using this algorithm. However, the performance of the RR algorithm is usually not as fast as other optimized algorithms. The following section gives an example of using the RR algorithm.

6.3.1.2 RR Algorithm Example

We give an example of using the RR algorithm in this section. Assume that there are three input applications and each application has a few tasks. One application is a BE application and the other two applications are AR applications. The arrival time of the three tasks is given. Moreover, there are three cloud servers, which are Cloud 1, Cloud 2, and Cloud 3. The execution time for tasks on each cloud server is given in Table 6.1. Figure 6.8 illustrates the DAGs of three applications and arrival time information.

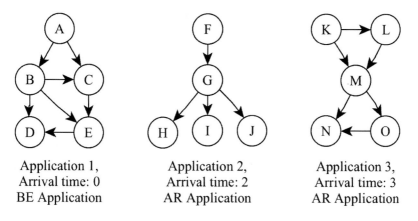

Application 1,
Arrival time: 0
BE Application

Application 2,
Arrival time: 2
AR Application

Application 3,
Arrival time: 3
AR Application

Figure 6.8 DAG for the motivational example. Three applications with 15 tasks in total. Each application has 5 tasks. Application 1 is a BE application. Applications 2 and 3 are AR applications. The arrival time information is given in the figure.

According to Figure 6.8, the tasks' execution sequence can be determined from the DAGs, which form the list P.

- Application 1: A, B, C, E, D.

- Application 2: F, G, (H, I, J). H, I, J tasks can be executed in a parallel manner.

Table 6.1 The Execution Time Table.

	A	B	C	D	E	F	G	H	I	J	K	L	M	N	O
Cloud 1	1	5	6	8	4	4	9	3	5	7	10	2	4	3	6
Cloud 2	3	9	4	12	5	3	8	4	4	5	7	4	4	1	8
Cloud 3	4	3	9	7	3	5	7	5	7	6	8	5	7	2	5

- Application 3: K, L, M, O, N.

Using the RR algorithm we can schedule the tasks in a plan that is given in Table 6.2.

Table 6.2 Task Assignment Using the Round-Robin Algorithm.

	A	B	C	E	D	F	G	H	I	J	K	L	M	O	N
Cloud	1	2	3	1	2	1	2	3	1	2	1	2	3	1	2
Time	1	9	9	4	12	4	8	5	5	5	10	4	7	6	1

Time	0-1	2-6	6-16	16-21			30-36	36-40	
Cloud 1	A	F	K	I	Idle		O	E	Idle
Time		1-6	6-14	14-19	19-23	23-27		36-37	40-52
Cloud 2	Idle	B	G	J	L	B		N	D
Time				14-19		23-30	30-39		
Cloud 3		Idle		H	Idle	M	C		Idle

Figure 6.9 The execution time and the execution orders for three clouds using the RR algorithm.

Figure 6.9 represents the results of using the RR algorithm. According to the figure, the total requested time is 52 ended by executing task D. Since Application 1 is a BE application, task B is suspended when Application 2's task G interrupts the execution of B. Task K starts at 6 even though Application 3's arrival time is 3, because Task K needs to be loaded in the waiting queue when another AR task F is in the execution. Furthermore, the cloud servers are in an *Idle* mode when there is no task assigned to them, which is also marked in the figure.

6.3.2 Cloud List Scheduling Algorithm

6.3.2.1 Cloud List Scheduling Algorithm Description

The *Cloud List Scheduling* (CLS) algorithm is an approach using the Greedy algorithm for scheduling the listed tasks in order to reduce the total execution cost, such as the execution time or energy consumption. Compared with the RR algorithm, the CLS algorithm chooses the cloud server having the highest value rather than a circular order. The algorithm mainly consists of two crucial steps, which are *List* and *Assignment*. At the list step, all input tasks are listed in a schedule using the Greedy algorithm. The cloud manager server always selects the available cloud server that requires the minimum cost. For instance, the manager server will choose the server requiring the shortest execution time for a certain input task if the system is designed to shorten the total execution time.

Next, the assignment will become active once the list is formed. The assignments will follow the list schedule such that the tasks will be assigned to those cloud servers that require the earliest finish time. This procedure will not terminate until the list is empty. A brief algorithm description is given as follows:

Require: a list schedule P using a priority-based method, the number of cloud servers m, and an ETM showing each task's cost at each cloud server.

Ensure: a static resource allocation plan.

1. Do the following operations **While** the list P is not empty (enter a While loop)

    (a) Pop out the first task T from the list P.

    (b) Pull cloud server information from all manager servers in order to find the earliest server available time for task T. The transferring time between servers can also be considered at this step, which depends on the system configurations.

    (c) Based on the information about the earliest server available time, find the minimum earliest estimated finish time of task T to which the task T is assigned. The task preemptions are not considered. Then remove task T from the list P.

2. Output the resource allocation plan.

An implementation example of the CLS algorithm is given in the following section.

6.3.2.2 Cloud List Scheduling Algorithm Example

We use the same DAGs used in Section 6.3.1.2. The DAG figures are given in Figure 6.8. The list P of CLS algorithm is as the same as the one used in RR algorithm. Table 6.3 represents the task assignment using the CLS algorithm. Figure 6.10 shows the execution time as well as the task execution sequence for three clouds. In this case, the total execution time is 36, which is much shorter than RR's total execution time, 52. The execution time is shortened by 30.77%.

Table 6.3 Task Assignment by Using the CLS Algorithm.

	A	B	C	E	D	F	G	H	I	J	K	L	M	O	N
Cloud	1	3	2	3	3	2	3	1	2	3	2	1	1	3	2
Time	1	3	4	3	7	3	7	3	4	6	7	2	4	5	1

Time	0-1				12-15	15-17	17-21			
Cloud 1	A	Idle		Idle	H	L	M		Idle	
Time			2-5	5-12	12-16	16-20			26-27	
Cloud 2	Idle	Idle	F	K	I	C	Idle	Idle	N	Idle
Time		1-4		5-12	12-18		20-21	21-26	26-28	28-35
Cloud 3	Idle	B	Idle	G	J	Idle	E	O	E	D

Figure 6.10 The execution time and the execution orders for three clouds by using the CLS algorithm. The bolded tasks are AR tasks.

Executing CLS algorithm does not consider the transfer time. The algorithm considering the transfer time is given in Section 6.3.3.

6.3.3 Min-Min Scheduling Algorithm

6.3.3.1 Min-Min Scheduling Algorithm Description

The *Min-Min Algorithm* (MMA) is a popular scheduling algorithm that is often combined with Heuristic or Greedy algorithms. A common MMA usually does not consider the precedence of the tasks, such as using a DAG. In cloud resource allocation, the task precedence is usually considered a prerequisite due to multiple tasks in an application. Therefore, we will introduce an advanced MMA algorithm considering transferring time and dynamic task dependence in this section.

This algorithm is called the *Min-Min Scheduling Algorithm* (MMSA), and derives from CLS algorithm. The algorithm also uses the Greedy algorithm to choose cloud servers, which is designed to reduce the total execution time while considering the transferring time. Two important parameters are considered in this algorithm, which include the earliest resource available time considering the transferring time response from other manager servers, and the earliest finish time of each task. The task list P is continuously updated in order to meet the tasks' dependences. The brief description of the algorithm is given as follows:

1. Initialize the task list P.

2. Execute the following operations for all tasks, **While** there are tasks awaiting the assignments (Enter a While loop).

3. Pull cloud resource status information from all available manager servers. Search the information according to the following two conditions that form the min-min mechanism.

   (a) Search **the earliest cloud resource available time** when considering the transferring time. This transferring time refers to the response time of the dataset from other manager servers.

   (b) Search and select **the earliest finish time** of the task from the earliest available time list. The preemptable tasks are not considered.

4. Assign the task to the selected cloud server. Remove the task from the list P after the task is assigned to a cloud server.

5. Update the task list P.

The *Estimated Finish Time* (EFT) is a sum of three time consumptions, including the cloud's *Earliest Available Time* (EAT), *Transferring Time* (TT), and *Execution Time* (ET). Assume that the task t_i is assigned to cloud j, and three parameters are $EAT_{i,j}$, $TT_{i,j}$, and $ET_{i,j}$. The mathematical expression is given in Equation (6.5).

$$EFT(t_i) = EAT_{i,j} + TT_{i,j} + ET_{i,j} \qquad (6.5)$$

The value of $TT_{i,j}$ can be found from the disk image size $S(t_i)$ and network bandwidth NB. An estimate method of calculating $TT_{i,j}$ is shown in Equation (6.6). The given formula obtains the transferring time by knowing the task volume and the data transmission efficiency.

$$TT_{i,j} = \frac{S(t_i)}{NB} \qquad (6.6)$$

In summary, MMSA is a resource allocation algorithm that is designed to shorten the execution time of the application in a dynamic cloud environment. The transferring time is considered by this algorithm. A simple example of MMSA is represented in the following section.

6.3.3.2 Min-Min Scheduling Algorithm Example

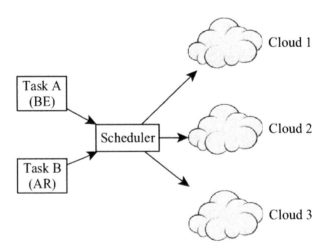

Figure 6.11 Two input tasks A and B. Three cloud server options, Cloud 1, Cloud 2, and Cloud 3.

We present an example to explain the mechanism of the MMSA algorithm. Assume that there are two input tasks, namely Task A and B. Task A is a BE task and Task B is an AR task. These two tasks have different arrival time spots: Task A's arrival time is 0 and Task B's arrival time is 7. There are three cloud servers in the system, Clouds 1, 2, and 3. The mission is to assign these two tasks to these cloud servers by using MMSA algorithms in order to obtain an efficient resource allocation schedule. Figure 6.11 represents the implementation scenario for the proposed example.

Table 6.4 shows the execution time table for both Tasks A and B. For example, Task A's earliest available time at Cloud 1 ($EAT_{A,1}$) is 2 and its earliest available time at Cloud 3 ($EAT_{A,3}$) is 5. The transferring time of Task A to Cloud 1 ($TT_{A,1}$) is 2. Meanwhile, Task B has the same earliest available time at all cloud servers because it is an AR task with the arrival time at 7.

Table 6.4 Execution Time Table for Tasks A and B*.

Task A				Task B			
Cloud	EAT	TT	ET	Cloud	EAT	TT	ET
1	2	2	12	1	8	2	8
2	3	1	9	2	8	1	6
3	5	2	4	3	8	2	4

* Tasks A's arrival time: 0; Task B's arrival time: 7.

The choice of the cloud server will depend on the earliest finish time for both Tasks A and B. The earliest finish time acquisition derives from Equation (6.5). For instance, Task A's earliest finish time at Cloud 1 is 16, which is gained from 2+2+12. Using this method can help us to draw the following Table 6.5.

Table 6.5 EFT Time Table for Tasks A and B*.

Task A		Task B	
Cloud	EFT	Cloud	EFT
1	16	1	18
2	13	2	15
3	11	3	14

* Tasks A's arrival time: 0; Task B's arrival time: 7.

According to the table, the smallest value of the EFT for Task A is given by Cloud 3, with 11 time units. For Task B, Cloud 3's EFT is 14, which is earlier than both Clouds 1 and 2. Therefore, both Task A and Task B will be assigned to Cloud 3.

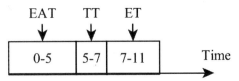

The estimated execution of task A at Cloud 3

Figure 6.12 The diagram of the estimated execution time for Task A at Cloud 3.

Moreover, calculating the total execution time of Task A needs to consider the interruption caused by Task B. Figure 6.12 illustrates a diagram of the estimated execution time for Task A at Cloud 3. The total estimated execution time is 11 and consists of three components. In the figure, 0-5 refers to the earliest available time for Task A at Cloud 3. 5-7 means 2-unit time for transferring time. And 7-11 means the execution time of Task A. This is a process that does not consider the preemptions.

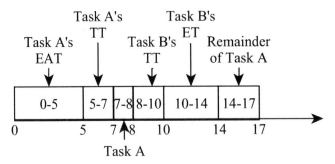

The actual execution of Task A when it is preempted by Task B at Cloud 3

Figure 6.13 The diagram of the estimated execution time for Task A at Cloud 3 while A is preempted by Task B.

In practice, BE tasks are often preempted by AR tasks. In this example, Task B preempts Task A at time 8 when Task B arrives. Figure

6.13 shows the actual execution time of Task A when the preemption takes place. As a BE task, Task A is split into two parts. The first part is operated from time 7 to 8. From time 8 on, Task A is suspended and Task B is inserted into the executions. The time 8-10 is the transferring time of Task B and it requires 4-unit time for the execution. Task A resumes execution after Task B ends at time 14. Then, the remainder of Task A is finished from 14 to 17.

In summary, MMSA is an approach using the Greedy algorithm to solve the dynamic problems in the cloud resource allocation mechanism. Two crucial parameters are considered in this algorithm, including the earliest cloud resource available time and the earliest finish time. Distinguishing from the CLS algorithm, MMSA counts the transferring time when calculating the total execution time. The cloud server j chooses for task t_i aims to select the shortest execution time that sums up $EAT_{i,j}$, $TT_{i,j}$, and $ET_{i,j}$.

6.4 FURTHER READING

For students who want to further deep this approach, further readings are available from reading the prior publication:

1. J. Li, M. Qiu, Z. Ming, G. Quan, X. Qin, and Z. Gu. Online optimization for scheduling preemptable tasks on IaaS cloud systems, *Journal of Parallel Distributed Computing, 72*(5), pp. 666–677, Elsevier, 2012.

2. K. Gai, L. Qiu, H. Zhao, and M. Qiu. Cost-aware multimedia data allocation for heterogeneous memory using genetic algorithm in cloud computing. *IEEE Transactions on Cloud Computing*, 2016.

3. M. Qiu and E. H.-M. Sha, Cost minimization while satisfying hard/soft timing constraints for heterogeneous embedded systems, *ACM Transactions on Design Automation of Electronic Systems (TODAES), 14*(2), pp. 1–30, Apr. 2009

4. Y. Li, M. Chen, W. Dai and M. Qiu, Energy-aware dynamic task scheduling for smartphones in mobile cloud computing, *IEEE System Journal*, 2015.

5. W. Dai, L. Qiu, A. Wu, and M. Qiu. Cloud infrastructure resource allocation for big data applications, *IEEE Transactions on Big Data*, 2016.

6. Y. Li, M. Qiu, W. Dai, and A. Vasilakos. Loop parallelism maximization for multimedia DSP in mobile vehicular clouds, *IEEE Transactions on Cloud Computing*, 2016.

6.5 SUMMARY

This chapter focuses on the preemptable algorithms in the optimizations of cloud computing. The main knowledge points of this chapter include:

1. Stated the background of the cloud optimizations in IaaS cloud systems.

2. Introduced the fundamental structure of the cloud resource allocation mechanism and the main steps of the mechanism. Four layers in the resource allocation mechanism are cloud users, Internet, manager server, and physical infrastructure. Four main steps included the partition, schedule, assignment, and execution.

3. Learned two messaging methods, which were pull and push modes.

4. Gave the concepts for the main items in cloud resource allocation models, such as AR and BE tasks; the method of formulating resource allocation problems; and the important parameters and variables in calculating the execution time.

5. Introduced three algorithms, including the *Round-Robin, Cloud List Scheduling* and *Min-Min Scheduling* algorithms.

6.6 EXERCISES

This exercise section provides students with an opportunity to practice three algorithms introduced in this chapter. The input applications include Application 1 and Application 2. As shown in Figure 6.14, Application 1 has 8 tasks, from Task A to Task H. Application 2 has 5 tasks, from Task I to M. Application 1 is a BE task and Application 2 is an AR task. Application 1's arrival time is 0. Application 2's arrival time is 7. Please complete the following exercises:

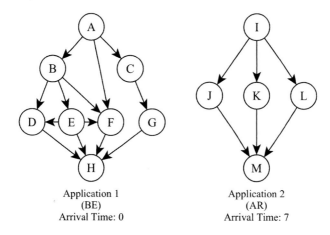

Application 1
(BE)
Arrival Time: 0

Application 2
(AR)
Arrival Time: 7

Figure 6.14 DAG for the exercise. Two applications: Application 1 and Application 2. Application 1 is a BE task; Application 2 is an AR task.

1. **Exercise for RR and CLS:**

    (a) Make lists for RR and CLS:

Table 6.6 The Execution Time Table.

	A	B	C	D	E	F	G	H	I	J	K	L	M
Cloud 1	2	4	7	9	4	4	11	3	4	7	10	6	4
Cloud 2	4	6	5	7	6	5	10	3	3	5	7	8	3
Cloud 3	4	2	8	8	5	6	9	5	2	6	8	9	7

Use the execution time shown in Table 6.6 to make a task assignment list by using the RR and CLS algorithms. Please complete Table 6.7 by filling in the cloud server selections.

Table 6.7 Task Assignment List.

	A	B	C	E	D	F	G	H	I	J	K	L	M
RR													
CLS													

(b) Complete Tables 6.8 and 6.9 by calculating the total execution time and using the list made in Table 6.7. Students can refer to Figure 6.9 and 6.10 as references.

Table 6.8 The Execution Time and the Execution Orders for Three Clouds Using the RR Algorithm.

Time	
Cloud 1	
Time	
Cloud 2	
Time	
Cloud 3	

Table 6.9 The Execution Time and the Execution Orders for Three Clouds by Using CLS Algorithm.

Time	
Cloud 1	
Time	
Cloud 2	
Time	
Cloud 3	

(c) The total execution time of using the RR and CLS algorithms:

RR:

CLS:

2. **Exercise for MMSA:**

This exercise also uses the same DAGs provided by Figure 6.14.

(a) Make a list for MMSA:

Use Tables 6.11 and 6.12 to make a task schedule list for MMSA. Fill in Table 6.10 by scheduling cloud servers.

(b) Calculate the total execution time:

Table 6.10 Task Assignment List.

	A	B	C	E	D	F	G	H	I	J	K	L	M
MMSA													

Table 6.11 Execution Time Table for Application 1.

Cloud	Task A			Task B			Task C		
	EAT	TT	ET	EAT	TT	ET	EAT	TT	ET
1	0	1	2	0	1	4	0	1	7
2	0	2	4	0	1	6	0	1	5
3	0	1	4	0	1	2	0	2	8

Cloud	Task D			Task E			Task F		
	EAT	TT	ET	EAT	TT	ET	EAT	TT	ET
1	0	1	9	0	2	4	0	1	4
2	0	1	7	0	2	6	0	1	5
3	0	1	8	0	1	5	0	1	6

Cloud	Task G			Task H		
	EAT	TT	ET	EAT	TT	ET
1	0	1	11	0	2	3
2	0	1	10	0	1	3
3	0	1	9	0	1	5

Table 6.12 Execution Time Table for Application 2*.

Cloud	Task I			Task J			Task K		
	EAT	TT	ET	EAT	TT	ET	EAT	TT	ET
1	7	1	2	0	1	4	0	2	7
2	7	1	4	0	2	6	0	1	5
3	7	1	4	0	1	2	0	2	8

Cloud	Task L			Task M		
	EAT	TT	ET	EAT	TT	ET
1	0	2	9	0	1	9
2	0	1	7	0	1	9
3	0	1	8	0	2	9

* The earliest available time of the Tasks J, K. L, and M depend on the finish time of their predecessors.

6.7 GLOSSARY

- **Cloud Manager Server** is a cloud resource management entity in cloud computing, by which a series of resource operations are accomplished, including resource information retrievals, task assignments, computation process displays, communication interconnections, and operation surveillances. This layer can be a physical infrastructure or software-based solutions using virtual operating interfaces.

- **Customers (in resource allocation mechanism)** are the cloud manager server that assigns an application/task to other cloud manager servers.

- **Entry Task** is the task that has no preceding task.

- **Execution Performance Matrix (EPM)** a technique that is applied for identifying the differences in execution performance between VMs, such as the execution time and energy cost.

- **Exit Task** is the task that has no succeeding task.

- **Preemptable Tasks** refers to an ongoing task that can be temporarily suspended when another input task having a higher-level priority is inserted into the task sequence and the task needs to be resumed after the inserted task finishes in cloud computing.

- **Producer (in resource allocation mechanism)** is the cloud manager server that receives the assignment of the application/-task from another cloud manager server, which is known as a *Customer*.

- **Pull Mode** allows customers gain information about the computing resource status from the producers in a *Pull* messaging mode.

- **Push Mode** allows producers to send/deliver the computing resource status information to customers when the resource status is updated in a *Push* messaging mode.

CHAPTER 7

Big Data and Service Computing in Cloud Computing

CONTENTS

Big Data AND SERVICE COMPUTING in cloud computing is an important aspect of service deliveries. This chapter focuses on these two concentrations: the first focus is big data and service computing in cloud computing; the other focus is phase-reconfigurable shuffle optimizations for MapReduce in cloud computing. Reading this chapter can assist students to not only have a basic picture of big data implementations in cloud computing, but also to learn the advanced algorithms of MapReduce in the cloud context. After reading this chapter, students will know the followings:

1. What is Big Data and how is it connected the concept of cloud computing?

2. What is the basic mechanism of MapReduce?

3. What is a *Phase-Reconfigurable Shuffle*?

4. Is it possible to optimize the phase-reconfigurable shuffle in Hadoop MapReduce?

7.1 INTRODUCTION

Big data is a technical term describing the techniques for retrieving information from large-sized data, which is usually used for generating values for large volumes of data storage or processing. The exact concept of big data is still not fully defined since it is currently at its exploring stage. However, there are three widely accepted characteristics of big data: the data are high in *Volume*, *Velocity*, and *Variety*. This is also known as the *V3s* standard. Understanding these characteristics is a fundamental step in understanding big data. The next level of understanding big data is to gain value from the large volume of data. The value retrievals are varied and depend on the data users' needs and targets.

Furthermore, many of today's organizations are seeking valuable information from a large volume of data to assist them in making a firm decision [45, 46, 47, 48, 49, 50, 51]. Normally, a lot of information can be retrieved from a large dataset. However, there are at least two main challenges in this issue. First, the data processing speed is a great concern because the target dataset has huge amounts of data. Second, efficient information filtering can also be a restriction since there will be a lot of *Noises* generated when data mining or data analysis is conducted. These challenges imply two major tasks of big data in computer science. One task is to speed up the computations when the data size is huge. The other task is finding out how to increase the accuracy of the information retrieval in order to let the gained information be meaningful and valuable to users. The proportion of the unexpected information needs to be removed from the discovered information.

Moreover, there are various frameworks, models, and techniques being used in big data. Currently, Hadoop MapReduce has become a leading open source framework for supporting big data growth and implementation [52, 53]. It is also a platform serving a giant size data processing or information retrievals. The operating principle of Hadoop MapReduce will be briefly reviewed. Based on the understanding of Hadoop MapReduce, an advanced algorithm using phase-reconfigurable shuffle is introduced in this chapter.

7.2 OVERVIEW OF BIG DATA

7.2.1 Concepts of Big Data

Using big data assumes that big data can reveal some information that cannot or hardly be acquired from traditional analysis techniques. As we mentioned in Section 7.1, there are three *V*s showing the main characteristics of big data. We give a structure image in Figure 7.1 and summarize these three *V*s as follows:

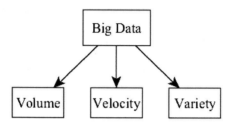

Figure 7.1 Three Vs structure of big data characteristics.

1. **Volume:** The first term, *Volume*, refers to the giant data quantity. Currently, data generations are much bigger than they were before. Data can be generated at every single second in a person's life, from movements to making calls. For example, Twitter, one of the social networks, normally generates more than 12 TB of data from more than 500 million tweets every day [54]. In essence, a large volume of data is a basic characteristic of big data.

2. **Velocity:** The term *Velocity* refers to a high computational speed. This characteristic is usually related to computing performances and practical usage. In many situations, users consider not only the accuracy of the output but also the efficiency of the data processing. For example, *machine-to-machine* data exchange can be done among billions of devices, in which the data exchange efficiency is a crucial part for satisfying real-time service demands.

3. **Variety:** *Variety* refers to the dramatic diversity of data types. This is also one of the most challenging dimensions in big data. The data can be either structured data, semi-structured or unstructured data, from numbers to video data. Most traditional database systems were built for limited volumes of structured data, which lacked predictable updatability. Compared with the traditional databases, the concept of big data addresses a much bigger mixture consisting of many types of data.

7.2.2 Big Data Processing

There are two main steps in big data processing: integrating data and using Hadoop MapReduce. A brief description of these two steps are given as follows:

1. **Integrating Data:** This step aims to integrate various data sources in preparation for using Hadoop MapReduce. It requires a number of procedures:

   (a) Map the data by extracting data features and finding out the connections.

   (b) Transforming the mapped data into a processable format and subdividing data in preparation for Hadoop MapReduce.

2. **Using Hadoop MapReduce**

   (a) Create the Hadoop MapReduce jobs followed by the outputs of the *Integrating Data* step.

   (b) Execute Hadoop MapReduce jobs and generate output.

Figure 7.2 illustrates a task flow diagram of MapReduce that is used for producing the ordered data from multiple data sources. In the figure, cubes with different pattern fills refer to different types of input data. The data are ordered and output after a series of operations, which are *Map, Shuffle,* and *Reduce.*

The main workload of big data is within Hadoop MapReduce. The advanced big data technique introduced in this chapter mainly focuses on Hadoop MapReduce. The optimizations are provisioned by improving the shuffle phase. The next section introduces the operating principle of the phase-reconfigurable shuffle optimization.

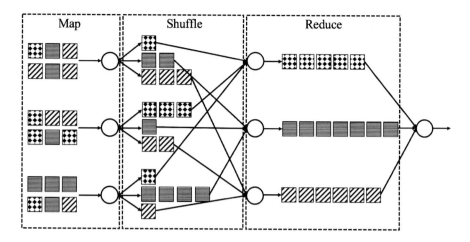

Figure 7.2   A diagram showing the operating mechanism of MapReduce.

7.3   PHASE-RECONFIGURABLE SHUFFLE OPTIMIZATION

As one of the popular approaches to big data, MapReduce's performance is still being restricted by a large number of parameters. Most parameters are attached to the *Shuffle* phase. The *Shuffle* phase refers to a group of procedures that are operated between Map and Reduce, in which sorting, grouping, and HTTP transferring are usually included. Shuffle is a time-consuming process on disk I/O that generally requires a long execution time, which results in an inefficient throughput. This restriction is caused by the complexity of Hadoop MapReduce's implementations, including both disk and networking activities. For example, copying a large amount of data to distributed remote nodes will make those nodes stop to wait for awaiting the data retrievals.

7.3.1   Spilling and Network in Shuffle

Normally, shuffle operations need to wait a long time for storing *Intermediate Result* (IR) files in MapTracker and sending IR to various ReducerTrackers in which the reduce functions are offered. An IR file refers to a type of data that is first written in local disks, which are available for the execution of the reduce functions. In Hadoop, IR files are distributed and assigned to the proper functional units (workers) for processing. This phase consumes around 1/3 of the execution time when using Hadoop MapReduce to manipulate big data. There are mainly three steps in a typical shuffle, which include a *Map-Side Shuffle, Networking*, and a *Reduce-Side Shuffle*. The main operations include:

1. Spilt or merge IR data.

2. Fetch IR data from different remote nodes.

Furthermore, Hadoop's implementations need enormous distributed disks that are supported by a *Hadoop Distributed File System* (HDFS). These distributed disks have two functionality offerings in MapReduce, involving data supplication and IR caching. A large input data file needs to be divided into a number of smaller data chunks. The divided smaller-sized data chunks will be sent to the remote MapTrackers after the MapTrackers are aroused. In addition, the IR shuffle process has not only disk operations but also the sorting or grouping phases. This data split-send process mainly consists of the following steps:

1. List the records of the location information for all data blocks in a file by sending a file inquiry to the name node.

2. Divide these data blocks into smaller-sized chunks. The local map function iteration will be initialized when MapReducers receive the divided data block chunks.

3. The MapTracker's IR will be concurrently added to a memory buffer when executing the mapping function. (There may be a number of split data files in the local disk that are waiting to be sent to the remote ReduceTracker. The reason is that the buffer contents will be temporarily split in local disk if the buffer is full.)

In addition, another time-consuming phase is *Data Transferring* in the distributed system. The IR need to be sent to remote nodes since MapTrackers and ReduceTrackers are located at the distributed nodes. Currently MapReduce executions barely consider the nodes' positions. This employment can result in a lower-level utilization of networks and inefficient data transferring. For example, some heavy workloads may suffer long-distance communications when light workloads take over the nodes nearby. Therefore, optimizing the data transferring mechanism is also an optimization alternative.

7.3.1.1 Map Functions in Distributed System

The first procedure of MapReduce is the *Map*, which has a number of steps. As we mentioned before, the Map procedure puts IR into a memory buffer until the buffer cannot accept more IR. Once this migration stops, a sorting operation on these buffered data will be executed by using a background thread in order to split the data and store them in the local disk. There will be a few split files stored in the disk having presorted IR data after the map function finishes. Moreover, a *Group* operation will be executed. These split files will be grouped to connect various files in which IR have the same key values. This means that the IR in the same group have the same key value. The process of the group can also enable a certain order of all IR in each group. Next, the groups will be sent to the remote ReduceTracker for further operations.

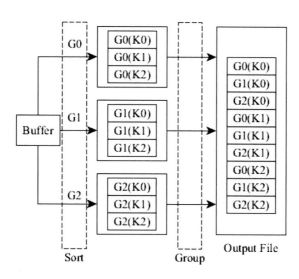

Figure 7.3 A sample of the map function operations.

Figure 7.3 represents an example showing the map function operations. In the given example shown in the figure, the MapTracker conducts and sorts 3 split files, which are G0, G1, and G2. In each file, there are a few IR data that have different key values, including K0, K1, and K2. Moreover, IR data are sorted according to the background threads. For instance, in G0, three IR data are sorted: G0(K0), G0(K1), and G0(K2). Furthermore, all split files will be put together by a *Group* function operation. This operation groups the IR data.

All IR data having the same key values will be grouped together. The output file is generated when all files are grouped.

7.3.1.2 Networks in Shuffle

Delivering the files to the target Reduce nodes is the next mission of MapReduce after the output files are generated. In most situations, the data transferring uses tree-type networks and HTTP connections. As illustrated in Figure 7.3, a Ki's IR data are gained from MapTrackers and are sent to the ReduceTrackers. This process can be time consuming because the networking performance will impact on the remote data transferring. Figure 7.4 shows a diagram of finishing time and network response for 280 MapTracker virtual nodes with WordCount benchmark.

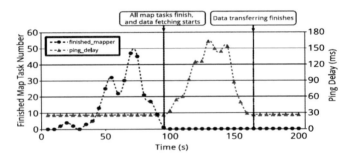

Figure 7.4 The diagrams of finishing time and network response for 280 MapTracker virtual nodes with WordCount benchmark [1].

7.3.1.3 Reduce Side

Figure 7.5 illustrates a sample structure of the Reduce function operations in shuffle. As shown in the figure, there are N output files generated by the *Map* functions. For example, assume that the Output File 1 is the output file of the Map function given in Figure 7.3. Moreover, there are M Reduce Tasks, thus the output files are sent to those reduce tasks for generating the final merged sequence. In the figure, we give a partial operation of Reduce Task 1, which is shown in the right-hand box. For example, the box shows that there are four output files with the grouped IR data. The IR data include G0(K0), G1(K0), G2(K0), G3(K0), G4(K0), G5(K0), G6(K0), G7(K0), G8(K0), G9(K0),

G10(K0), and G11(K0). Eventually, a single sequence is generated by merging the four grouped IR data.

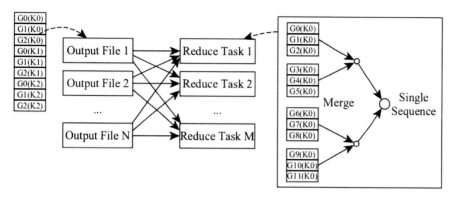

Figure 7.5 A sample structure of the network function operations in shuffle.

7.4 FURTHER READING

For students who want to research this approach in depth, further readings are available from reading the prior publication:

1. J. Wang, M. Qiu, B. Guo, and Z. Zong, Phase-reconfigurable shuffle optimization for Hadoop MapReduce, *"IEEE Transactions on Cloud Computing"*, vol. PP, no. 99, pp. 1–13, IEEE, 2015.

2. P. Zhang, L. Ling, Y. Deng. A data-driven paradigm for mapping problems, *Parallel Computing*, vol. 48, pp. 108–124, 2015.

3. K. Gai, M. Qiu, L. Chen, and M. Liu, Electronic health record error prevention approach using ontology in big data, in *The IEEE 17th International Conference on High Performance Computing and Communications (HPCC)*, New York, NY, August 24–26, 2015, pp. 752–757.

4. H. Yin, K. Gai, and Z. Wang, A classification algorithm based on ensemble feature selections for imbalanced-class dataset, in *The IEEE International Conference on High Performance and Smart Computing (HPSC)*, pp. 245–249, New York, NY, April 2016.

5. P. Zhang, Y. Gao, M. Qiu. A Data-oriented method for scheduling dependent tasks on high-density multi-GPU ystems, in *The IEEE 17th International Conference on High Performance Computing and Communications (HPCC)*, New York, NY, August 24–26, 2015, pp. 694–699.

6. H. Yin and K. Gai. An empirical study on preprocessing high-dimensional class-imbalanced data for classification, in *The IEEE International Conference on Big Data Security on Cloud*, New York, NY, August 24–26, 2015, pp.1314–1319.

7.5 SUMMARY

This chapter focuses on big data and service computing. The main knowledge points of this chapter include:

1. The basic concept of big data is introduced in this chapter. Three major characteristics of big data include volume, velocity, and variety.

2. Typical big data processing has two steps, which include integrating data and using Hadoop MapReduce.

3. We have introduced and reviewed the main mechanisms of MapReduce.

7.6 EXERCISES

1. What are the meanings of *Volume*, *Velocity*, and *Variety* in the concept of big data?

2. Briefly describe two steps of big data processing.

3. Briefly describe the mechanism of phase-reconfigurable shuffle optimization taught in this section. Explain the operating principle of the optimizations.

4. What is an *Intermediate Result* (IR) in Hadoop?

5. Give a presentation on the spilling and network in shuffle.

6. How does a *Map* work in a distributed system?

7.7 GLOSSARY

- **Big Data** refers to a technical term describing the techniques for retrieving information from large sized data, which is usually used for generating values for large quantity of data storage or processing.

- **Intermediate Results** refers to a type of data that are first written in local disks, and are available for the executions of the reduce functions.

- **Shuffle** a group of procedures that are operated between Map and Reduce, in which sorting, grouping, and HTTP transferring are usually included.

- **Variety** (in big data) refers to the dramatic diversity of data types.

- **Velocity** (in big data) refers to a high computational speed.

- **Volume** (in big data) mainly refers to the giant data quantity.

III

Security Issues and Solutions in Mobile Cloud Systems

CHAPTER 8

Security and Privacy Issues and Threats in Mobile Cloud Computing

CONTENTS

Security AND PRIVACY are significant aspects for mobile cloud users, developers, and service vendors, since any privacy leak may result in serious unexpected consequences. Currently, users' sensitive information is facing various threats, some of which are caused by the implementations of Web or mobility technologies. Many security problems in mobile cloud computing are generally associated with privacy issues. In this chapter, we summarize and review major security and privacy problems and introduce the main taxonomy of threats in mobile cloud computing. Students will be able to answer the following questions after reading this chapter:

1. What are security and privacy in mobile cloud computing? What are their features?

2. What is the problem of *Loss of Control* in cloud computing?

3. Why can *Multi-Tenancy* cause security/privacy problems?

4. What are the risks of *Massive Data Mining*?

5. What does an *Attack Interface* mean?

6. What is a *Threat Model* and why is it important for discerning a security/privacy problem?

7. What are the differences between an insider threat and an outside threat?

8. What are two sides in insider threats?

8.1 INTRODUCTION

In this section, we introduce a few basic concepts of security and privacy issues in mobile cloud computing, as well as the main features of the issues.

8.1.1 Basic Concepts

Along with the emergence of mobile cloud computing, security and privacy are two great concerns for most implementations due to numerous adversaries and threats. Many individuals like talking about security and privacy together. However, security and privacy are two distinct issues that have different concentrations, even though they are two closely related technical terms. The relations and differences between security and privacy are still a controversial topic that does not reach a consensus [46, 55]. We define these two concepts as follows, which will be used throughout this book.

Security in mobile cloud computing, also known as *Data Security*, refers to any mechanism that defends the mobile cloud systems against any adversaries for any purposes of abusing, breaching, or damaging data during the the entire cloud service delivery process. This issue emphasizes the approaches of securing data, including all related security-oriented methods, tools, strategies, mechanisms, framework, and operations. In general, security issues and solutions are highly considered by service vendors. A security issue is normally a data-related matter that does not consider much about the information carried by the data. For securing a system, cloud system designers need to enhance the security level to protect all data from any unexpected and improper behaviors. One of the purposes of increasing security is protecting privacy. A few aspects of security are covered by this issue, which will be introduced in the following sections.

Next, the *Privacy* issue in mobile cloud computing mainly focuses on sensitive information leakage during the mobile cloud service processes. The sensitive information can be owned by either mobile users/customers or service venders. The task scope of privacy covers data protections, secure operations, and proper data maintenance. Generally speaking, privacy issues are important concerns for customers. Privacy has a tight relationship with security as the security vulnerabilities can potentially result in the privacy leakage, due to the unexpected information disclosure caused by data security weaknesses. For example, abusing patient data on the cloud-based tele-health system can accidentally or purposely release patients' sensitive information, such as medical records and movement history.

Figure 8.1 represents the relations between security and privacy. There are some overlaps between the two entities that cover both data

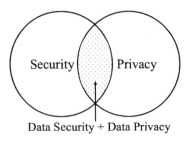

Figure 8.1 Relations between security and privacy.

security and data privacy. The following section introduces a number of features of security and privacy issues in mobile cloud computing.

8.1.2 Features of Security and Privacy Issues in Mobile Clouds

Despite security and privacy issues, mobile clouds have many similarities to non-mobile cloud computing, and there are a few distinguishing features. This section summarizes two main features of security and privacy issues in mobile clouds.

First, a security or privacy solution is generally associated with one single or a group of similar security or privacy problems, rather than a large scope of problems [55, 56]. The challenging part in this issue is that a security problem exists at all layers during the service delivery process in mobile cloud systems. It is difficult for system designers to talk about security and privacy without considering specific problems or cases. New threats can be created by new technologies or methods, which might be developed for a certain application scenario [57, 58]. Many attacks are also unanticipated or evolutionary. Therefore, it is almost impossible to use one particular method to defend the system against all adversaries. A good security and privacy strategy is to establish a secure framework with a high-level data protection view, as well as consider all potential security risks based on the existing/known threats [59].

The other feature is the inheritance of the vulnerabilities. The vulnerabilities are normally attached to the technologies used in mobile cloud computing. As described in Chapter 3, three main technologies include mobile computing, wireless networks, and cloud computing. The vulnerabilities of mobile cloud computing inherit all vulnerabilities of these technologies. For example, the vulnerabilities that exist in wireless networks also threaten the implementations of mobile clouds, such as spoofing and jamming attacks [60, 21]. From the perspective of cloud computing, implementing virtualization techniques is a unique characteristic in cloud computing, since the virtualization is re-shaping the relationships between the *Operating Systems* (OS) and connected hardware or infrastructure, including mobile ends and wireless networks. The ample use of this technique brings a brand new virtualization layer to the system, in which many cloud service configurations are formed, such as task assignment/distributions and cloud resource management. Thus, whether or not the virtualization layer is secure is a concern because operation failures of the virtualization layer can result in severe consequences to the cloud systems.

The following section introduces the main problems of security and privacy in mobile cloud computing.

8.2   MAIN SECURITY AND PRIVACY PROBLEMS

In this section, we concentrate on the contemporary major problems of security and privacy in mobile cloud systems. Our primary goal is to understand the common challenges from the following aspects, which are information over-collection, data or privacy control, trust management, and multi-tenancy problems. These aspects emphasize the cause of the security problems. The purpose of discerning these aspects is to understand where the security problems stem from, so that the solutions or improvements can be made by considering vulnerabilities.

8.2.1   Data Over-Collection Problems

Traditional security problems are playing a less important role in the mobility context, such as malware. The major goal of attacking mobile cloud computing is to obtain un-permitted information. The *Data Over-Collection* (DOC) problem is an emerging threat that endangers mobile users' privacy. This new threat commonly exists in mobile applications [61, 62, 63]. A large number of mobile applications carry a large volume of data containing sensitive information, which enlarges the attack surface throughout the mobile cloud system. The outcomes of the DOC problem can cause other security and privacy problems as well, such as data leakage on the third party.

The background of this issue is mostly twofold: First, the broad usage of mobile applications has provided a great number of opportunities for collecting users' data on various mobile platforms. The data storage on the clouds can be distributed and one single mobile app allows multiple cloud servers to support its services. Currently, there are millions of mobile apps being active. Each mobile app collects a bunch of data for the purpose of service offerings, such that a pool of attack objectives is formed. Second, the difficulty of decrypting data is remarkably high with the development of communication protocols and cryptography [12, 64]. This phenomenon motivates attackers to find out other methods of gaining information from the unencrypted data. Many unencrypted data used by mobile apps have become targets.

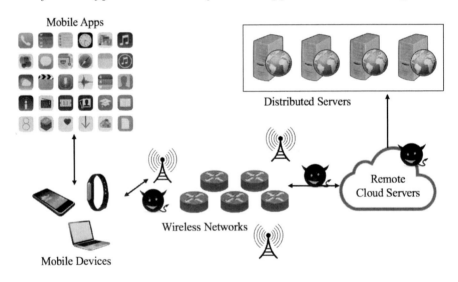

Figure 8.2 Data over-collection in mobile cloud computing.

Figure 8.2 shows a few attack targets for DOC problems in mobile cloud computing. The devil icons represent the adversarial target locations. Mobile users' data are collected by different service providers via VM and wireless networks. It implies that many data duplications carrying mobile users' privacy will be distributed through various channels. A great number of channels provide adversaries with more opportunities. The risks exist in either the data transmission process or the operations on the cloud side, such as data processing and data storage [21].

Next, Figure 8.3 further explains the cause of data over-collections. As shown in the figure, assume that there exists a mobile app offering

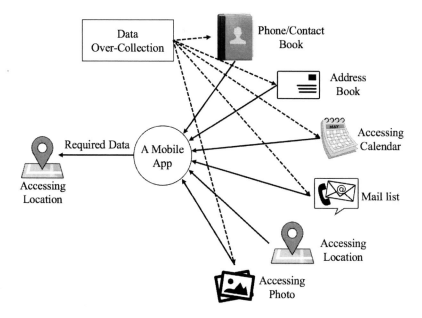

Figure 8.3 Illustration of the data over-collection problem.

location-oriented services. For the purpose of business analysis, this mobile app collects not only location data but also other data, including phone/contact data, address data, calendar data, mail list data, and photo data. As a matter of fact, only location-oriented data are related to its service offerings. All other data collections belong to the scope of over-collections. This problem is also an example of loss of control from the customer side.

8.2.2 Data/Privacy Control Problems

The problem of data/privacy control is mainly caused by using remote computing resources. This concern is mainly on the customer side. Cloud customers are not supposed to have full control of data, applications, or other computing resources located with service providers. The access or system control configurations are determined by system designers or administrators. Customers usually lack sufficient authority even though they realize the potential threats, in some situations. Thus, cloud customers highly rely on the secure operations processed by cloud service providers.

Moreover, user identity management is operated and maintained by the cloud operators. This means that the user access control rules and security policies and enforcement are defined by cloud providers. The cloud side governs all authentication and verification operations, rather than sharing any control authority with the customer side.

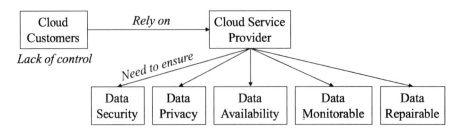

Figure 8.4 Dependence diagram showing the main aspects that cloud service providers need to ensure.

Due to the high dependence relations, cloud service providers need to ensure a few aspects of data in order to guarantee the safety of customers' data security and privacy [21]. This is also a common concern when system developers design a new cloud system. Figure 8.4 shows a dependence diagram that displays five aspects, including security, privacy, availability, monitoring, and repair. All these five aspects can be the target of the adversary. Data security and data privacy are two primary and fundamental concerns in ensuring proper operations of cloud systems.

Furthermore, *Availability* means whether the data or relevant cloud services are available when the service requests are sent out by cloud customers. Next, *Monitoring* refers to whether the data stored or governed by the clouds can be tracked by the customer who is also the data owner. The tracking authority levels can be varied. Finally, *Repair* emphasizes that the data can be repaired when the cloud system suffers an overwhelming disaster or data breach. More details will be given in Section 8.3.1.

8.2.3 Trust Management Problems

In the cloud-based context, *Trust Management* (TM) is a type of *Information Technology* (IT) management model that uses conceptual systems to represent, determine, and secure data security and privacy within a social-based operating environment, which is usually designed for assisting cloud system developers in making firm decisions on information security, policy generation, and trust assessment. Consider the characteristics of mobile cloud systems, TM issues are often attached to third-party issues, such as intercrossed cloud platforms and switchable cloud services [65, 58, 66, 67].

From both customer and vendor sides, there are two major concerns with respect to third-parties. First, we need to understand the method of estimating whether the third party is trustable. This concern can also be stated as: *How can we trust a third party?*. Next, cloud system designers need to have a good picture of the entire system and avoid unnecessary resource offloading to the third party. The level of risk will be enhanced if the number of the third parties increases, even though most third parties can be considered trustworthy. The adversarial activities may occur in various scenarios, such as *Virtual Machine* (VM) mitigations and physical machine sharings. Therefore, the system developers had better make a proper decision on whether to take a risk.

Furthermore, we have many methods of improving TM in mobile cloud systems. Most methods follow two principles: understanding the trust issues and the existing/potential risks and correctly defining the problems based on the understandings. For example, in social networking TM, understanding the behaviors of each object in the networking system is a critical component for building reliable and secure trustworthiness management [68, 69].

8.2.4 Multi-Tenancy Problems

Multi-tenancy is one of the characteristics of cloud computing that also brings a few new problems. The origin of multi-tenancy problems is that multiple users share the same physical machine. Figure 8.5 illustrates the situation of the same physical machine shared by multiple users. As shown in the figure, both normal users and attackers have services that are given by the same physical machine. This implies that the attackers can legitimately located in the server with other normal users. A similar problem in *Infrastructure-as-a-Service* (IaaS) is that adversaries map the cloud infrastructure due to the legitimacy of attackers in the system.

A few conflicts can result in this challenge. For example, conflicts may exist between cloud tenants when a pool of computing resources are shared by tenants. In this case, the tenants may have different

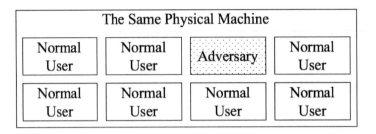

Figure 8.5 Multi-tenancy issues in the cloud: normal users and adversaries share the same physical machine.

or opposing business goals or service demands, such that the required computing resources could be various. In order to meet different demands, the solution needs to cover two aspects, compatibility and isolation. The *Compatibility* in multiple-tenancy issues refers to whether various services can be delivered by using the same physical machine. The *Isolation* in multiple-tenancy issues means if one single physical machine can support the customer and the service individually. The challenge is designing the method of separating services.

8.2.5 Summary

In summary, this section briefly talked about the main problems of security and privacy in mobile cloud computing. Four common aspects in this field are summarized. Being aware of these mentioned problems and general causes is the fundamental issue of defining or finding the security or privacy problems, which is a necessary condition for solving a specific problem.

8.3 THREAT TAXONOMY

In line with the sections above, we have a quick review of the threat taxonomy of security and privacy in mobile cloud computing in this section. Despite many taxonomy dimensions in this field, we mainly talk about five primary threat taxonomies, which include confidentiality, integrity, and availability, massive data mining, attack interfaces, auditability and forensics, and legal issues. Similar to the prior sections, the main purposes of understanding threat taxonomy are twofold. First, understanding threat taxonomy is a fundamental requirement of conceptualizing security or privacy problems. Second, developing a security or privacy solution needs to be aligned with threats. Figure 8.6 represents a graphical mapping of threat taxonomy, which is addressed in this section.

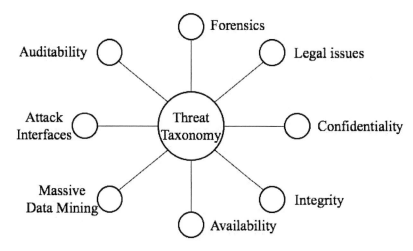

Figure 8.6 Threat taxonomy mapping.

8.3.1 Confidentiality, Integrity, and Availability

In cloud computing, *Confidentiality* refers to ensuring that customer information stored on cloud servers is not accessed by unauthorized persons using data analysis, data integrity, or data mining techniques. The retrieved data can be from either a single cloud server or a group of distributed cloud servers. In the cloud-based context, the main concern is that the cloud vendors have full operational authority. Cloud customers have confidentiality concerns when their sensitive data are stored in the cloud. Whether or not the cloud vendors' operations are sufficiently secure is a critical consideration [70, 71].

Integrity refers to integrating multiple cloud systems that can ensure that all cloud customers can only access the services or data that they are authorized to access. The challenges of integrity mainly derive from performance concerns [55]. Cloud system designers need to ensure that the system can process a job effectively and the system can output an accurate job.

Finally, *Availability* issues in clouds refer to whether the cloud vendors can guarantee that services can continuously be offered to customers. Service discontinuity and distortion can occur for a few reasons. First, cloud services can be suspended or disrupted when service providers are attacked, such as *Denial of a Service* (DoS) [46]. Second, cloud services can be terminated when the service providers quit the business due to various reasons, such as financial burdens or technical restrictions. Finally, service availability can be restricted if the service provider cannot provide sufficient service scalability. The limited service scope will result in reduced service availability.

8.3.2 Massive Data Mining

Privacy risks caused by *Massive Data Mining* (MDM) mainly derive from using data mining algorithms to obtain sensitive information from the large volume of data. This risk can occur in different situations. In some cases, the adversaries obtain data from multiple cloud platforms using data mining techniques. In some other situations, adversaries can successfully map the cloud infrastructure and monitor the data on a cloud server in which multiple cloud customers' data are stored. Data mining algorithms offer a great chance to breach customers' privacy, even though only the partial information is stored in one source or some data are encrypted.

8.3.3 Attack Interfaces

In mobile cloud computing, attacks can happen at different layers since the attack targets are various. Malicious attacks can also occur at the communication links due to insecure interfaces and *Application Programming Interfaces* (APIs). The main threat is that the insecure API cannot provide effective authentication capabilities by which the authentication, system control, and authorization can be protected from API hacks. A variety of threats are attached to attack interfaces, such as confidentiality, availability, and integrity.

8.3.4 Auditability, Forensics, and Legal Issues

Both auditability and forensics are data control issues. Auditability and forensics are difficult because the data are stored remotely rather than being stored locally in the mobile cloud context. In order to ensure that the operations of auditability and forensics can be successfully carried out, a secure framework, strategy, plan, or model is required in terms of business processes and needs [59, 46]. *Mobile Cloud Auditability* (MCA) refers to a set of activities that support regulatory compliances in terms of the agreements between mobile cloud customers and vendors.

Mobile Cloud Forensics (MCF) refers to a group of techniques designed for identifying, assessing, gathering, and analyzing cloud data during the process of information retrieval and behavior supervision or governance in mobile cloud computing. MCF can be considered a subset of *Cloud Forensics*, which is associated with *Digital Forensics* [72]. From the technical perspective, a few activities need to be addressed, including gathering data and analysis, forensics status examinations, evidence assessment, proactive forensics strategy, and virtualization technique alternatives. Figure 8.7 represents the relationships among MCF, cloud forensics, and digital forensics.

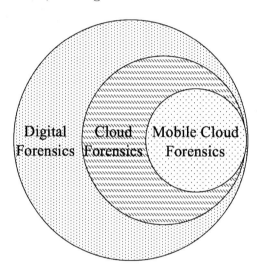

Figure 8.7 Relationship chart for MCF, cloud forensics, and digital forensics.

Legal issues in this field are dynamic due to the continuously emerging laws and regulations. Many aspects are involved in this dimension, since service providers need to ensure that their services and cloud-based behaviors follow the law or corresponding regulations. Considering the technical part, legal issues in mobile cloud computing emphasize the significance of identifying the party who is in charge of this issue. In some cases, confusion can be caused by the messy relations of the third parties, since cloud service providers may subcontract to other parties.

8.3.5 Summary

In summary, we had a brief review on threat taxonomy of mobile cloud computing in this section. A few definitions are given that will be used throughout this book. The next section introduces a few common threat models in the mobile cloud context.

8.4 THREAT MODELS

In this section, we concentrate on the concept of threat models and present two types of threats, including insider and outsider threats. Section 8.4.1 introduces the basic concepts related to the threat model. The following sections provide detailed descriptions of the threat categories in mobile cloud computing.

8.4.1 Basic Concepts

A *Threat Model* is an approach that is designed to formulate the security problem by accomplishing a serious of tasks, including assisting in analyzing security problems, designing a mitigation strategy, and proposing solutions. Forming a threat model is an important step for developing a security solution. The target threat needs to be clearly defined and identified in the threat model. Figure 8.8 illustrates a workflow diagram showing the main steps of developing a threat model. Normally, building a threat model consists of a few steps as follows:

1. The first step is that cloud system designers or security experts need identify the target threats. Defining a threat requires understanding a group of factors, such as attackers, adversary model, malicious activities, threat occasions, threat objectives, and other relevant factors. In a research problem, the scope of the target threats should not be too broad in order to keep the problem within a solvable range.

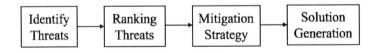

Figure 8.8 Workflow diagram of developing a threat model.

2. The next step is ranking threats in terms of the outcomes of the first step. The threat rankings need to be aligned with the identified factors, which can be configured as either parameters or variables depending on the requirements of security problems.

3. The third step is choosing a mitigation strategy. In mobile cloud computing, risk mitigation is a crucial part of determining the primary strategy for security solutions. The specific security layer needs to be addressed at this step.

4. Finally, a solution to the target threats is developed.

Moreover, at the step of identifying/defining a threat, the system designers should be aware of the attacker or adversary model. The *Attacker/Adversary Model* (AAM) refers to a descriptive model providing the threat information, including determining the attacker types, analyzing attacker motivations, and understanding attacker capabilities. There are a variety of approaches to categorizing attacker types. For example, the attacker types can be grouped into insider attackers and outsider attackers. The other method of grouping attacker types is to distinguish single attackers from collaborative attackers. According to statistics from *Breach Level Index* (BLI), Figures 8.9 and 8.12 show the distributions of data breach incidents. Figures 8.9, 8.10, and 8.12 show the data breach incidents source collected in 2014, 2015, and 2016, respectively. Figure 8.11 shows data breach incident reported by industry in 2015. In this book, we mainly discuss the differences and implementations of insider and outsider threats, addressed in Sections 8.4.2 and 8.4.3.

In general, it matters for cloud customers to know the trust levels of each service. From the perspective of cloud deployment, a public cloud does not provide a higher-level security guarantee; thus its trust level is normally lower than a private cloud. In a common situation, cloud customers assume that cloud insiders are not adversaries. However, in some research problems, this assumption is also a vulnerability of cloud-based applications. Some examples of the potential problems caused by insider attacks include: (1) cloud operators gain customers' information without permission; (2) cloud operators know customer behaviors by using data mining techniques; (3) customer information is sold to a third party.

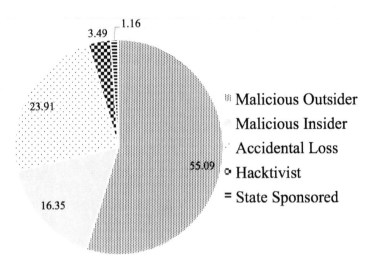

Figure 8.9 Distributions of data breach incidents by source for 01/01/2014–12/31/2014. Data retrieved from breachlevelindex.com.

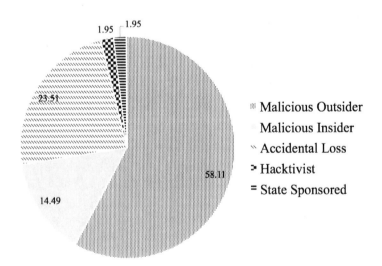

Figure 8.10 Distributions of data breach incidents by source for 01/01/2015–12/31/2015. Data retrieved from breachlevelindex.com.

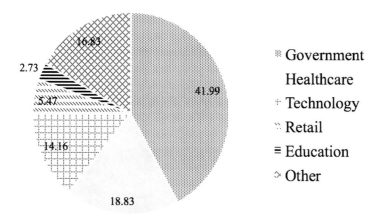

Figure 8.11 Distributions of data breach incidents reported by industry for 01/01/2015–12/31/2015. Data retrieved from breachlevelindex.com.

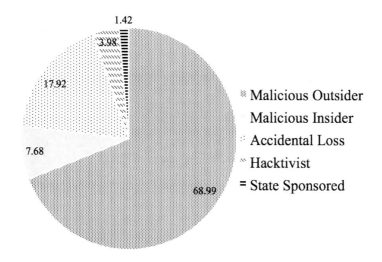

Figure 8.12 Distributions of data breach incidents by source for 01/01/2016–06/01/2016. Data retrieved from breachlevelindex.com.

8.4.2 Insider Threats

In this section, we mainly discuss one category of the threat model, which separates threats into insider threats and outsider threats. *Insiders* are parties that have the authority or privilege to access the full or partial data stored on cloud servers. The *Insider Threat*, also known as the *Malicious Insider*, refers to any risky activities or improper operations occur at the one of the insider parties in the mobile cloud system, which is either the client side or the service provider side. Figure 8.13 is a structural graph showing the crucial in insider threats.

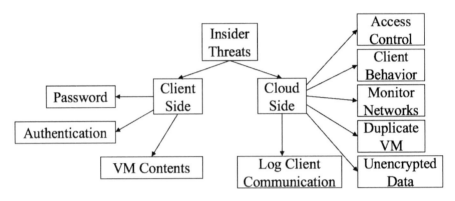

Figure 8.13 Mapping insider threats.

8.4.2.1 Customer Side

As shown in the figure, there are a few potential risks on the client side. First, malicious clients intend to obtain other users' password information via adversarial activities. Using similar methods can also enable malicious clients to obtain other information that they are not supposed to have, such as other clients' authentication and verification information. Moreover, some adversarial clients target the contents of VMs or mapping cloud infrastructure. The effects of malicious activities triggered by clients can result in various consequences. Some of them are serious, such as account information leakage in the financial industry. The challenge is great for service providers because the attack methods, channels, and occasions are varied and unanticipated. It is hard to deal with attackers who pretend to be normal customers.

The essential mechanism of diminishing risks is to reduce the attack surface. The attack chance will be enhanced when any distribution applications are employed. In most situations, system designers should avoid unnecessary distributions, since the security level of distribution-based applications is usually lower than the one held by a less-distributed application, such as establishing the system within a

Local Area Network (LAN). Another approach is to utilize proxy and brokerage services to avoid cloud clients directly accessing the shared physical machines.

8.4.2.2 Service Provider Side

The other threat type in insider threats derives from the service provider side. In real-world cases, this concern is usually not involved in the consideration elements when making a mobile cloud strategy. Service providers always claim that their operations are fully secure. Pointing out this threat in a business negotiation can simply destroy the efforts made by providers and result in collaboration failures. However, the threat does exist and the issue is an open research topic in academia. The main feature of insider threats from the cloud side is that cloud operators or employees have access to or authority over cloud customer data. Related to this feature, some common threats from cloud operators or employees include logging client communications; reading customers' unencrypted data without permissions; duplicating VM containing customers' private data; monitoring networks and recording transmissions; obtaining customers' behaviors, habits, and routes by using data mining techniques; and stealing account information by using access control governance authority.

To address this issue, a number of methods are available to enhance the security level. In order to prevent insider threats from improper operations made by cloud employees, security administrators need to ensure they have sufficient surveillance on activities occurring on the system. Having a tracking mechanism is an option for recording insider operators' activities. However, this method is not a proactive approach and is mainly used to periodically assess and surveil the system rather than a real-time system monitoring. The record system generally creates a large volume of data, which needs regular elimination to save physical storage space.

Furthermore, another alternative solution for the cloud-side insider threats is using *Service-Level Agreements* (SLAs) to negotiate the aspects related to data security and privacy, such as logging authority, auditing mechanism, and regulatory compliance. An SLA is a contract that is used to define specific cloud services by clarifying all required aspects, such as service scopes, qualities, and responsibilities. An SLA is usually a component of *Standardized Service Contract* (SSC). The following section addresses the other type of threat, which is the outsider threat.

8.4.3 Outsider Threats

The *Outsider Threat* is any malicious activities launched by the parties that are unauthorized and outside the mobile cloud computing systems. The primary motivation of outsider threats is to obtain the protected information. In general, an outsider does not have credentials and is not authorized to have direct access to the system.

8.4.3.1 Attack Methods and Goals

The continuous improvement in Web technologies has enabled attackers to develop a variety of attack methods and a variety of attack goals. We summarize a few common attack methods as follows:

1. *Malware Injection Attacks:* There are many metadata exchanges between web servers and web browsers in Web services when cloud solutions are implemented. *Malware-Injection Attacks* (MIAs), also known as *Metadata Spoofing Attacks* (MSAs), refers to a group of adversarial methods that invade the cloud systems by injecting malicious programs. MIAs often take place when the metadata are exchanged. The malware codes can be injected by attackers who pretend to be the valid users in the clouds. A successful malware injection attack can cause a few risks to cloud systems, such as communication eavesdropping, operation deadlocks, and peeping cloud mapping structure.

2. *Inserting Malicious Traffic:* Due to the requirement of wireless networks, inserting malicious traffic can result in a great impact on the system's performance. Separating the malicious packets from legitimate packets is a great challenge in wireless networking traffic. Internet traffic behaviors are dynamic and variable so that formulating the unusual behaviors is not an easy task. Most current solutions use monitoring techniques that surveil traffic behaviors, such as *Intrusion Detection Systems* (IDSs) [73]. The methods of malicious traffic are varied, such as DDoS and Worm attacks.

   Furthermore, *Denial-of-Service* (DoS) in mobile cloud computing refers to those adversarial actions that try to prevent mobile users from reaching the cloud computing resources. The cloud services can be suspended, interrupted, or terminated due to the adversarial attack. Currently, in cloud systems, a common DoS attack is *Distributed DoS* (DDoS), which uses distributed attack source and shares attack objectives. The attack can address at least two types of computing resources by using DDoS attacks, which are networking bandwidth-oriented and computing resource-oriented attacks [74]. In the networking bandwidth-oriented DDoS, flooding attacks and amplification attacks are two common attack approaches. Protocol vulnerability exploitations and sending malformed packets are two basic attack methods of computing resource-oriented DDoS.

3. *Probing Network Traffic:* Distinguished from inserting malicious traffics, probing networking traffic is an attack method that concentrates on the networking capacity estimation of the ad hoc wireless networks in mobile clouds. Like other attack types, many attack methods are available, depending on the attack focus and technique. For example, in one of methods of probing network traffic, the critical objective is estimating bottleneck capacities [25], even though the there are many other parameters for assessing path capacities, such as topology, interference, and antennae. The attacker aims at vulnerabilities of bottlenecks in order to interfere with normal communications.

4. *Side-Channel Attacks:* A *Side-Channel Attack* (SCA) refers to any adversarial activity that uses information derived from the cryptosystem's physical implementations instead of directly invading the system. The scope of the information in this domain covers any information leaked from the cloud system In the mobile cloud context, for instance, adversaries may impersonate an authorized user in the cloud-based platform by collecting sensitive application data from other co-located users and obtaining sufficient information to hijack the target user's account [75]. This approach is an extension of the *Flush-Reload Attack* (FRA) mechanism, which is relatively devoid of noise [76, 77].

8.4.3.2   Crucial Issues in Outsider Threats

From the perspective of attackers, the major challenge is co-locating itself with the attack objective in the same physical machine when launching a MIA. It is critical but hard for attackers to find the correct target locations. The network and remote server configurations can be continuously changed in the cloud context. Accurately locating vulnerabilities is a tough job in such a dynamic operating environment.

Moreover, *Software-as-a-Service* (SaaS) can provide a higher level of security than *Infrastructure-as-a-Service* (IaaS) and *Platform-as-a-Service* (PaaS), because SaaS allows service providers to ensure security as part of the SLA. The customer side in SaaS has less security responsibility than both IaaS and PaaS. Figure 8.14 represents the security responsibilities of cloud customers and service providers for IaaS, PaaS, and SaaS.

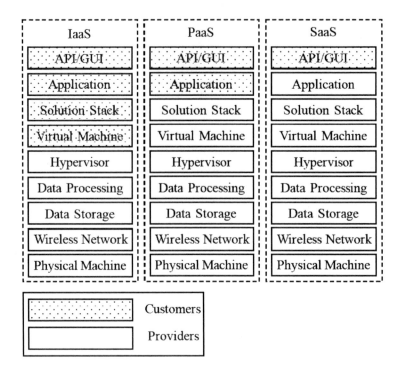

Figure 8.14 Security responsibility of cloud customers and service providers from the perspective of service models. IaaS: Infrastructure-as-a-Service; PaaS: Platform-as-a-Service; SaaS: Software-as-a-Service.

Additionally, as shown in the figure, there is a clearly defined security boundary in PaaS. API, *Graphic User Interface* (GUI), and applications are covered by customers. The service provider side is responsible for the rest of the parts, most likely including the software framework and middleware layer. IaaS provides a less secure level considering the greater responsibility taken by the customer side.

8.5 FURTHER READING

1. K. Gai, M. Qiu, M. Chen, and H. Zhao. SA-EAST: Security-aware efficient data transmission for ITS in mobile heterogeneous cloud computing, *ACM Transactions on Embedded Computing Systems (TECS)*, vol. 16, no. 2, pp. 6, ACM, 2016.

2. Y. Li, K. Gai, L. Qiu, M. Qiu, and H. Zhao. Intelligent cryptography approach for secure distributed big data storage in cloud computing, *Information Sciences (INS)*, vol. 387, pp. 103–115, Elsevier, 2016.

3. K. Gai, M. Qiu, X. Sun and H. Zhao. Security and Privacy Issues: A Survey on FinTech. In proceedings of *The International Conference on Smart Computing and Communication (SmartCom 2016)*, pp. 236–247, Shenzhen, China, Springer, Dec. 2016.

4. K. Gai, M. Qiu, H. Zhao, and W. Dai. Privacy-preserving adaptive multi-channel communications under timing constraints. In proceedings of *The IEEE International Conference on Smart Cloud (SmartCloud 2016)*, pp. 190–195, New York, USA, IEEE, Nov. 2016.

5. M. Qiu, K. Gai, B. Thuraisinghamb, L. Tao, and H. Zhao, Proactive user-centric secure data scheme using attribute-based semantic access controls for mobile clouds in financial industry, *Future Generation Computer Systems (FGCS)*, Elsevier, 2016.

6. K. Gai, M. Qiu, B. Thuraisingham, and L. Tao. Proactive attribute-based secure data schema for mobile cloud in financial industry. In proceedings of *The IEEE International Symposium on Big Data Security on Cloud*, pages 1332–1337, New York, USA, IEEE, 2015.

7. K. Gai, M. Qiu, and H. Hassan. Secure cyber incident analytics framework using Monte Carlo simulations for financial cybersecurity insurance in cloud computing, *Concurrency and Computation: Practice and Experience*, Wiley, 2016.

8.6 SUMMARY

In this chapter, we have covered a number of crucial aspects of security and privacy issues in mobile cloud computing. First, we introduced basic concepts of security and privacy, as well as their features. We gave the definitions for both security and privacy that are used throughout this book. Security refers to any mechanism that defends the mobile cloud systems against any adversaries for any purposes of abusing, breaching, or damaging data during the entire cloud service delivery process. Privacy focuses on sensitive information leakage during mobile cloud service processes. Second, we have summarized the main security and privacy problems. The main summarized problems included information over-collections, data/privacy controls, trust management concerns, and multi-tenancy problems. Third, we have addressed threat taxonomy by giving a few terms in the security and privacy field. The provided terms include confidentiality, integrity, availability, massive data mining, attack interfaces, auditability, forensics, and legal issues. Finally, we emphasized the significance of understanding threat models and presented two major threats, which were insider threats and outsider threats.

8.7 EXERCISES

1. Briefly describe the difference(s) between security and privacy in mobile cloud computing. What are their relationships?

2. List two features of security and privacy issues in mobile cloud systems and explain.

3. What hazards can be caused by the *Data Over-Collection* (DOC) problem in mobile clouds?

4. Explain why multi-tenancy of cloud computing can bring risks.

5. Describe the differences between *Confidentiality*, *Integrity*, and *Availability* in threat taxonomy.

6. What is *Massive Data Mining* (MDM)?

7. Briefly describe the relationships among *Digital Forensics*, *Cloud Forensics*, and *Mobile Cloud Forensics?*.

8. Jenny owns a company and she intends to introduce a mobile

cloud-based solution to improve her company's business services. The company focuses on providing data storage service offerings. This year, Jenny wants to expand the company's market shares by introducing mobile cloud services for other mobile app developers. The new service will provide extended data storage for mobile app service providers. The physical storage facility is distributed throughout America in order to ensure high performance in data retrievals.

In order to reduce the risk of using mobile clouds, Jenny decides to develop a *Threat Model* to analyze and predict the potential hazards. Picture yourself as a *Chief Information Officer* (CIO) in Jenny's company, and propose a detailed analysis and results of developing a threat model for the company.

8.8   GLOSSARY

Attacker/Adversary Model refers to a descriptive model providing threat information, including determining the attacker types, analyzing attacker motivations, and understanding attacker capabilities.

Availability issues in clouds refer to whether the cloud vendors can guarantee the services can be continuously offered to the customers (in security and privacy).

Compatibility refers to whether various services can be delivered using the same physical machine (in multiple-tenancy issues).

Confidentiality refers to ensuring that customer information stored on cloud servers is not accessed by unauthorized persons using data analysis, data integrity, or data mining techniques.

Insider is any party that has authority or privilege to access the full or partial data stored on cloud servers (in insider threat).

Insider Threat refers to any risky activities or improper operations occurring at one of the insider parties in the mobile cloud system, which is either on the client side or on the service provider side (in mobile cloud computing).

Integrity refers to integrating multiple cloud systems that can ensure all cloud customers can only access to the services or data that they are authorized to access (in security and privacy).

Isolation in multiple-tenancy issues means that one single physical machine can support the customer the service individually (in multiple-tenancy issues).

Mobile Cloud Auditability refers to a set of activities that support regulatory compliance in terms of the agreements between mobile cloud customers and vendors.

Mobile Cloud Forensics refers to a group of techniques designed for identifying, assessing, gathering, and analyzing cloud data during the process of information retrieval and behavior supervision or governance in mobile cloud computing.

Outsider Threat refers to any malicious activities launched by parties that are outside the mobile cloud computing systems (in mobile cloud computing).

Privacy focuses on sensitive information leakage during the mobile cloud service processes (in mobile cloud computing).

Security refers to any mechanism that defends the mobile cloud systems against any adversaries for any purposes of abusing, breaching, or damaging data during the entire cloud service delivery process (in mobile cloud computing).

Service-Level Agreement is a contract that is used to define specific cloud services by clarifying all required aspects, such as service scopes, qualities, and responsibilities.

Side-Channel Attack refers to any adversarial activity that uses the information derived from the cryptosystem's physical implementations instead of directly invading the system.

Threat Model means an approach that is designed to formulate the security problem by accomplishing a series of tasks, including assisting in analyzing security problems, designing mitigation strategy, and evaluating solutions.

Trust Management is a type of *Information Technology* (IT) management model that uses conceptual systems to represent, determine, and secure the data and privacy within a social-based operating environment, which is usually designed for assisting cloud system developers in making firm decisions on information security, policy generation, and trust assessment.

CHAPTER 9

Privacy Protection Techniques in Mobile Cloud Computing

CONTENTS

Techniques FOR PROTECTING SECURITY AND PRIVACY are critical aspects in securing services and operations in mobile cloud computing. In Chapter 8, we talked about basic concepts in security and privacy issues and threat models. Students should have an overview picture of security and privacy in mobile cloud computing after reading Chapter 8. This chapter focuses on a number of crucial security dimensions to assist students in understanding specific security problems. A few security solutions will be introduced in this chapter, too. For the purpose of a deep understanding of security solutions, we will introduce some advanced security algorithms at the end of this chapter.

Students will be able to answer the following questions after reading this chapter:

1. What are the main layers of infrastructure security?

2. What are the crucial issues in mobile data security and storage?

3. What are the major privacy concerns in mobile cloud computing?

9.1 INTRODUCTION

As mentioned in Chapter 8, security and privacy issues exist everywhere throughout the service process in mobile cloud computing. Basic security and privacy issues and threat models were introduced in Chapter 8. For further understanding of the specific security and privacy issues, we need to look into each dimension/layer and understand the particular reasons causing the risks. In this chapter, we primarily address three aspects of security and privacy issues in order to assist students in understanding the issues in detail. The main contents of three aspects are given as follows.

The first aspect is having a deep understanding of the security and privacy issues from four crucial dimensions. First, we will concentrate on the dimension of infrastructure security in mobile cloud computing. A few layers will be addressed in this dimension, including network, host, and application layers. For each layer, a description of risks and causes will be provided. Second, we will discuss the dimension of data security and storage in mobile cloud computing. A number of concepts regarding data processing and storage in clouds will be introduced in this part. Third, we will talk about identity and access management in mobile cloud computing. Finally, we will analyze data storage, data retention, and data destruction.

Next, we will focus on the aspect of security solutions. A few solutions will be summarized in terms of weaknesses in mobile cloud computing. Three perspectives are covered in this part, which are trust, control, and multi-tenancy. These three perspectives are also aligned with the vulnerabilities that were already addressed in Chapter 8. A few solutions are described based on the existing frailties in mobile clouds.

Finally, the next section will introduce an advanced security algorithm. In this part, the method of modelizing security problems and designing algorithms will be presented. Moreover, an algorithm designed for increasing security levels is described, which uses dynamic programming to obtain optimal solutions to balancing security and performance.

The following section centers the crucial security dimensions.

9.2   CRUCIAL SECURITY DIMENSIONS

9.2.1   Infrastructure Security

Infrastructure is a crucial security dimension in mobile cloud computing, because infrastructure is deployed throughout the entire mobile cloud system, from end users to remote cloud servers, from wireless networks to databases [73, 78, 5]. Therefore, we had better set up a variety of layers in the infrastructure security dimension in order to distinguish each risk from the large volume of threats. Figure 9.1 represents three principal aspects in infrastructure security. The following subsections give descriptions of these three aspects one by one.

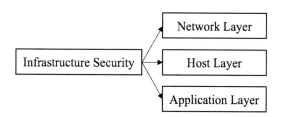

Figure 9.1   Three layers of infrastructure security in mobile cloud computing.

9.2.1.1   Network Layer

The *Network Layer*, also called the *Network Level*, refers to the security and privacy issues that exist or occur during the data transmissions on wireless networks. The ideal networking configuration is securing all data on the network, even though there exist a lot of vulnerabilities in today's wireless networks. At the same time, a variety of network traffic protection/encryption techniques have been developed, for instance, *Secure Socket Layer* (SSL) and the *Transport Layer Security* (TLS). Current ongoing network secure techniques need to consider most of the risks or threats mentioned in Chapter 8, such as confidentiality, integrity, authentication, authorization, and auditing. Three design principles concerning the security of the network are presented here:

1. The wireless network needs to guarantee good confidentiality and integrity for all *Data-in-Transit* (DiT) when cloud service offerings are provisioned.

2. The wireless network needs to guarantee secure, reliable, and efficient access control operations, covering authentication, authorization, and auditing.

3. The wireless network needs to guarantee a high level of computing resource availability, which includes both resources for cloud customers and the resources assigned to tenants from cloud service providers.

In essence, neither security risks nor risk levels are associated with service models. But it matters what kind of service deployment is used considering the security and privacy issues in mobile cloud computing. At the network layer, the selection of the network type can impact the risk types and levels. For example, private networks are normally more secure than public networks.

In general, encryption is an effective alternative for protecting DiT from data leakage. For some enterprises requiring strong security protection, all data on the networks are encrypted. However, the time or energy consumption may become an issue when the volume of data is large. Some data are unencrypted or partially encrypted in many situations. When data are in transit across networks, many adversarial opportunities are created, since attackers can obtain sensitive information from either monitoring wireless communication or breaking into the physical infrastructure. On an Ethernet network, adversaries may be able to steal sensitive information by overhearing a cable or mirroring network traffic by maliciously configuring a switch. Compared with the malicious attacks on Ethernet networks, adversaries usually use the vulnerabilities of wireless networks. Securing DiT needs to consider two aspects. The first side is that all involved nodes/hosts need to be secured. The other side is that the data transmissions/communications between nodes/hosts need to be secured.

For example, in the *Wireless Fidelity* (Wi-Fi) network environment, a variety of security protection protocols and programs are available, such as *Wi-Fi Protected Access II Personal* (WPA2-Personal), *Wi-Fi Protected Access II Enterprise* (WPA2-Enterprise), and *Wi-Fi Protected Setup* (WPS). Among these security protocols, WPA2-Personal, also known as WPA2-PSK (Pre-Shared Key), is a security protocol that is designed for home users or small organizations and only uses a single password. Next, WPA2-Enterprise is an enterprise-oriented security protocol that requires usernames and the corresponding passwords, such that one user can have multiple accounts. Finally, WPS is an emerging security standard for establishing a secure wireless home network. Generally, WPS has four working modes, including *PIN Method* (PINM), *Push Button Method* (PBM) [79], *Near Field Communication Method* (NFCM) [80, 81], and *USB Method* (USBM).

Box 9.2.1.1: About Data-in-Transit, Data-at-Rest, and Data-in-Use

Data-in-Transit (DiT) means all data flowing across both trusted and untrusted networks, which usually take place between two nodes on the network.

Data-at-Rest (DaR) means the data state when the data are stored/recorded on the cloud-based physical server/media in mobile cloud computing.

Data-in-Use (DiU) means the data state when the data are attached to one node in a network rather than the DaR state. Some examples of the particular nodes in a network include the processor cache, disk cache, and memory.

Furthermore, distinguished from DiT, there are two other common data states, which are *Data-at-Rest* (DaR) and *Data-in-Use* (DiU) [82]. The definitions of these three data states are given in Box 9.2.1.1. In order to further illustrate the differences among these data states, Figure 9.2 shows the locations of three data states on the network, which are aligned with the definitions represented in Box 9.2.1.1. Data at the DaR state can be protected by strong encryption without sharing the key with any nodes. The length of the key used at the DaR state is must be long enough. Data at the DiU state can be secured when access to the memory is fully under control and data can be gained at the rest state rather than any other locations.

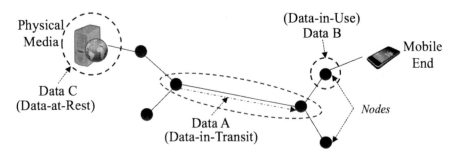

Figure 9.2 Illustration of Data-in-Transit, Data-at-Rest, and Data-in-Use.

Moreover, using *Virtual Machine* (VM) techniques has brought a variety of new networking deployments and risks to the network layer. The rise of public cloud computing has replaced or changed the traditional network zones and tiers. Network isolation is not necessarily attached to physical networks, since virtual networks can be deployed in public cloud computing and isolation of the networks can be achieved by domains. Figure 9.3 represents an example of a diagram showing the mechanism of virtualizing networks.

9.2.1.2 Host Layer

The security of the host layer concerns both local and remote hosts. SaaS and PaaS are two service models primarily considered at this layer. Host security is often considered the responsibility of service providers, since the host *Operating Systems* (OS) are abstracted and hidden from end users. This configuration leads to the assumption that cloud service providers should protect data security and privacy for mobile cloud customers. As a matter of fact, customers also own the risk of information protection and the local host also needs to be protected. Both local and remote hosts need to be secured.

The security risks on remote hosts are similar to traditional security issues on networking infrastructure. Data protection mechanisms are mature and effective in preventing risks. In addition, there are more

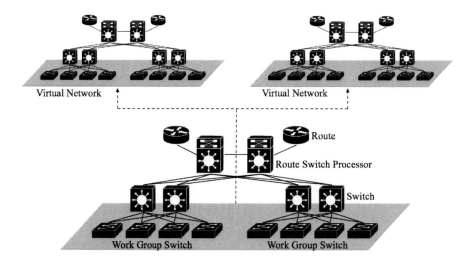

Figure 9.3 Diagram of virtualizing networks. The figure is adapted from Cisco [2].

threats at the host layer in mobile cloud computing because of a lack of security at local hosts. A *Local Host* (LH) refers to the mobile devices used by the mobile end users. The threats become stronger at LH due to a few reasons:

1. Threats caused by phone theft problems are serious. The statistics provided by *Consumer Reports* indicated that 3.1 million American consumers were victims of smartphone theft in 2013 [83]. This number was almost two times bigger than it was in 2012. Adversaries can obtain sensitive information directly from mobile devices and can gain authorized access to mobile cloud systems via the stolen phones.

2. Lost mobile devices can cause the same problems as mobile device thefts.

3. Sometimes, attackers launch buffer-overflow attacks when the weaknesses of mobile embedded systems are addressed [84, 85]. This type of attack applies to both special-purpose embedded systems or general-purpose mobile systems.

   A general solution to this problem is building up boundary checks, which are based on an assumption that all developers address this issue when a mobile embedded system is developed.

However, the remarkably increasing amount of mobile embedded systems has made this solution unreliable. The threat level is not reduced by applying this solution. Any tiny vulnerabilities may result in serious consequences, since mobile devices are connected to wireless networks such that they are authorized users having authenticated access to the cloud systems. Mapping cloud infrastructure is one of the threats caused by buffer-overflow attacks that occur at local hosts.

4. Mobile devices generally cannot provide the same security protection capability as on-premises devices, due to the restrictions of the device capability and communication configurations.

5. The vulnerabilities in wireless networks are the genes of mobile cloud computing. The weakness of wireless communication provides more attack channels for adversaries, such as monitoring emitted signals and fooling mobile devices with *Man-in-the-Middle* attacks.

In order to strengthen the local host security, there are a few improvement dimensions. First, authentication methods attached to the local host should be strengthened [58, 86]. Second, the tamper-resistant mechanism is one of the options for preventing attacks [87]. Third, the security level can be enhanced by ensuring strict application isolation [73]. Fourth, the trust level of the operating system at the local host can be increased [65, 88]. Finally, using the cryptographic approach is also an alternative for strengthening the local host security.

9.2.1.3 Application Layer

In the mobile cloud computing context, customers are mobile users who receive cloud services and use mobile devices as the local hosts. In many situations, the system is insecure even though the network layer is secure. Some risks derive from the application layer in which unauthorized users have access to the permission-required information. It is important for customers to understand the party who is responsible for application security and what the responsibilities are.

The security strategies may be various due to the different cloud service models. One type of strategy is to adjust integration and functionality in terms of the service models. For example, the security strategy of *Infrastructure-as-a-Service* (IaaS) is to reduce the level of the integrated functionality and security. *Software-as-a-Service* (SaaS) uses the security strategy that maximizes the integration level of the security. The major purpose of these strategies is to minimize the impacts caused by loss of control from the customer side. Some other security strategies are combining the application layer with other layers' security strategies, such as the host layer. A few current attacks can be effectively prevented by using this approach, such as *Cookie Poisoning* [89], *Hidden Field Manipulation Hacks* [90], *DoS Attacks* [91], and *Google Hacking* [92].

9.2.2 Mobile Data Security and Storage

Mobile data security and storage are associated with the data state, such as DiT and DaR. The concepts of DiT, DaR, and DiU are given in Section 9.2.1. In this section, we mainly talk about the security strategies or methods for different data states. For DiT, the data security and privacy levels can be categorized by protocols, including secured protocols and un-secured protocols. At the DaR state, data are generally non-encrypted and mixed with other users' data, since the encrypted data are not indexing/searching friendly. Currently, the implementations of processing data can be done by operating non-encrypted data. Potential solutions include *Fully Homomorphic Encryption* (FHE) and *Predicate Encryption* (PE), both of which are immature for industrial deployments.

Moreover, there are a few active compliance standards available for securing data storage, which are required by the governmental regulatory framework [20]. Most of these standards have specific focuses. Some examples of compliance standards include The *Payment Card Industry Data Security Standard* (PCI-DSS), The *Health Insurance Portability and Accountability Act* (HIPAA), and The *Gramm-Leach-Bliley Act* (GLBA). Following compliance standards is a major approach for securing data storage from the legal perspective. Figure 9.4 represents a few common acts of the governmental regulatory security framework, as well as brief descriptions.

In addition, in data usage security, there are two important aspects:

- *Data Lineage* refers to an approach defining a data life cycle that shows when and where the data are located. An effective data lineage is important for data usage security, such as data auditing and compliance.

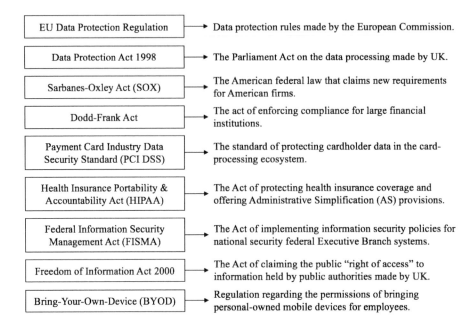

Figure 9.4 A few common acts of the governmental regulatory security framework.

- *Data Provenance* is a term describing computational accuracy when data integrity is operated. The meaning of data can be messed up if an ineffective data provenance is configured. An example is given in Box 9.2.2.1.

Box 9.2.2.1: Data Provenance

Question:
Determine whether the following expression is correct:
$\text{sum}((((2\times4)+6)/2)+3)=\$10$

1. It is correct if data are in US dollars.
2. It is incorrect if data are in different countries' currencies.

Next, cloud vendors usually have data storage service offerings by providing a package of services with servers. Two common approaches are available in terms of the level of storage administration, which are *Managed Cloud Storage* (MCS) and *Unmanaged Cloud Storage* (UCS).

The concepts of these two types of cloud storage are given in Box 9.2.2.2.

Box 9.2.2.2: Managed vs. Unmanaged Cloud Storage.

Managed Cloud Storage (MCS) is a type of data storage deployment that provisions raw disk services either on-site or off-site so that users can have authorization to configure the disk, such as partitions and formats. It is a type of on-demand online storage.

Unmanaged Cloud Storage (UCS) is a type of data storage deployment that provisions preconfigured storage so that users do not have authorization to configure the disk, such as formatting or installing file systems and changing drive properties. This type of data storage is normally more reliable than MCS.

9.2.3 Mobile Identity and Access Management

In mobile cloud computing, one solution is to develop mobile identity and access management in order to improve the security level. Three aspects can be addressed when designing access management, which are presented in Box 9.2.3.

Box 9.2.3: Mobile Identity and Access Management Dimensions.

User Groups: Mobile users can be grouped into a number of divisions. A division has a unique identity and access authentication for its involved users. For example, users in a company can be grouped as employers, employees, contractors, partners, etc.

Demands of Authentications: Increasing demands of authentication is another dimension of improving mobile identity and access management. A few new approaches include using cloud offerings to host applications, such as cloud-side authentication. This approach is usually called an "Outsider" solution that offers authentication in clouds.

Mobile Users: Identity and access management can also be improved by configuring mobile computing, including increasing security levels on mobile devices and mobile communications. This dimension concentrates on the mobile users.

9.2.4 Privacy Protection Concerns

9.2.4.1 Data Life Cycle

The term *Privacy* may have different meanings in various contexts, such as nations, cultures, and regulations/laws. To address privacy issues, properly understanding privacy within the environment is a fundamental requirement. For security professionals, the understanding of privacy can derive from discerning public expectations as well as legal statements. A few common rights/obligations related to privacy include data collection, usage, disclosure, storage, destruction, etc.

In most application scenarios, privacy protection is a matter in data life cycles. Figure 9.5 illustrates a data life cycle showing the process from data generation to data destruction. Each circle represents a phase in the data life cycle. Considering security and privacy issues, each phase can be treated as a concern dimension in most situations. Data security and privacy issues exist throughout the entire process of this life cycle.

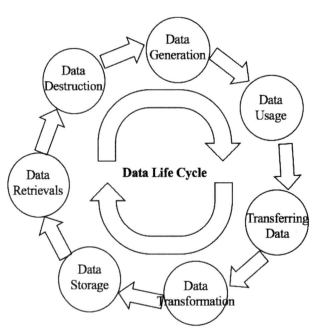

Figure 9.5 A common data life cycle.

The next section will introduce a few common concern dimensions in mobile clouds.

9.2.4.2 Concern Dimensions

In this section, we concentrate on introducing a few key concern dimensions in privacy issues in the mobile cloud context. The presented concern dimensions broadly exist in contemporary implementations.

Data Storage Dimension The privacy issues in the data storage dimension mainly derive from the cloud side, since most data are stored in remote cloud servers in mobile cloud computing. This feature leads to the fact that data share the same physical machines even though cloud users are distinct. On the cloud side, the data indexing and searching operations will not notify data owners in most implementations.

Moreover, a few behaviors of cloud service providers are often related with privacy issues. Sometimes, it matters to officially authorize whether the cloud service providers have the authentications to read data, even though most service providers have the access. A few methods of obtaining private information on the cloud side include using data to track users' behaviors, movements, and routes, acquiring information by launching data mining techniques, and trading information without notifying data owners.

Data Retention Dimension In the cloud context, the concept of data retention refers to the period length of storing data in clouds. The major issues of data retention considering privacy include:

1. Period length of keeping data in clouds.
2. Data ownership identification.
3. Responsibility clarification for enforcing policies.

Data Destruction Dimension Data destruction is generally done when data retention ends. In the mobile cloud context, data destructions are associated with both local and remote sides. The privacy issues related to this domain mainly focus on three aspects, including methods, results, and duplications. The first aspect is the method of data destruction. An improper destruction method can result in the hazards of data leakage. Next, whether or not the data are fully destroyed is an issue in protecting privacy. Data recovery techniques can help attackers obtain private information by recovering data that has not been completely destroyed. Finally, many cloud-based solutions offer distributed storage so that data can be available in distinct clouds. Data duplications can lead to the problem of privacy leakage when additional data copies are made in multiple data centers and the physical data storage locations are various. Data governance is challenging when multiple parties are involved in the service delivery process.

Data Auditing/Monitoring Dimension Data auditing/monitoring issues are associated with business process improvements and management. Identifying the steps of the business processes is the fundamental requirement. From the perspective of the participants in the business process, dimensions can be addressed in different ways when concentrations vary.

First, there are two crucial aspects of mobile users' perceptions in this filed: data auditing/monitoring methods and frequency. Being aware of these two aspects can assist mobile cloud users reduce concerns about privacy leakage. Moreover, on the cloud service provider side, a few tasks need to be addressed. First, the service provider needs to always have a plan to deal with an accidental cyber event. Second, the auditing/monitoring approach needs to consider the implementation scenario when multiple parties are involved. Finally, forming effective regulations of auditing/monitoring is a common approach.

Privacy Breaches Dimension Privacy breaches are usually considered one of the major concerns for mobile users when mobile apps are applied, since the adversarial results have a direct relationship to mobile users' benefits. Cloud service providers and cloud customers have different concerns in this dimension.

Cloud service providers may need to work on the following aspects, including discerning the frequencies and occasions of privacy breach attacks, preparing solutions for each type of breach attack, and formulating a mapping system for notifying customers about breach risks. Next, cloud customers have the right to understand the relevant regulations and the party who is in charge of preventing privacy breaches.

Data Responsibility and Risk Management Dimension Data responsibility and risk management is a new concern dimension that is usually related to multiple parties' participation in cloud computing. Figuring out the responsibilities and formulating roles in risk management are challenging when a third party is added between cloud service providers and customers. Lack of full reliance on the third party often results in the privacy issue, since it is difficult to define and restrict each role's responsibility. For most organizations or stakeholders, an alternative approach is to transfer liability, even though the accountability can rarely transferred. Furthermore, risk management is a popular method of modeling processes that articulates each role's tasks at each phase. Three basic components of formulating risk management include the model creation, model validation, and model process.

In summary, these concern dimensions are only popular concerns existing in current implementations of mobile cloud systems. There are many other concerns that mostly depend on the demands and applications. The next section will introduce a few mechanisms for solving security and privacy problems.

9.3 SECURITY AND PRIVACY SOLUTIONS

This section presents a few solutions to security and privacy issues mentioned in the prior section.

9.3.1 Overview

The solutions mainly derive from the weaknesses of the security and privacy issues. In line with the introduction before, security solutions usually address three perspectives, including trust, control, and multi-tenancy. These three aspects are associated with three features of cloud computing. Again, it is almost impossible to list all security and privacy threats due to the complexity of the relationships between participators in mobile clouds and the large variety of applied techniques. A basic mechanism of increasing the level of security and privacy protections is to minimize these three perspectives.

Box 9.3.1 summarizes the main security issues in mobile clouds for solution developers to address. The three issues summarized here include confidentiality, malicious behaviors, and emerging attacks/vulnerabilities.

Box 9.3.1: Security Issues

Confidentiality Issues: The confidentiality issues in mobile clouds mainly arise when "multi-" implementations occur, such as multi-tenancy, multi-service-provider, and multi-parties for data/information trading.

Malicious Behaviors: Malicious behaviors refer to those activities for improper purposes within the data usage process. Improper behaviors addressing adversarial activities can be launched by either insiders or outsiders.

Emerging Attacks/Vulnerabilities: New threats are mostly associated with the new features of cloud computing. For example, sharing data between *Virtual Machines* (VMs) can bring adversarial opportunities; collaborative attacks take place at multiple layers of the network.

9.3.2 Monitoring-Based Solutions

Monitoring-Based Solutions (MBS) for protecting privacy primarily increase the level of data control by improving data governance and observation. This approach offers a pool of benefits that address both cloud customers and service providers.

First, from the perspective of customers, using MBS can help cloud users have a better understanding of failure effects. The understanding process can enable cloud customers to know the recovery measurements when threats happen. Being aware of the steps of the business process can also help cloud customers have different views of the system for understanding potential risks. On the other hand, from the perspective of cloud providers, using an MBS mechanism can create an approach of re-mapping cloud infrastructure in order to deal with accidental attacks. An alternative is to turn off partial functions or sub-systems to reduce the loss caused by attacks. This method needs full awareness of the business process such that shutting down partial functions/sub-systems will not result in an unaffordable loss. Meanwhile, applying an MBS mechanism is also good for bringing systems under control. Activities within the process are surveilled, which provides fundamentals for forecasting, preventing, or reducing hazards.

Moreover, the implementations of MBS can be used to develop solutions in both occurring attacks and before an attack occurs. When considering the solution to an ongoing attack, MBS can support or create a number of aspects, such as designing access controls based on the monitoring results, partitioning/isolating physical machines in terms of monitored actions, or moving data to other cloud servers according to the monitored status. Meanwhile, using MBS can also create solutions before an attack. The collected data concerning the cloud systems' conditions and statuses can be used for creating back-up plans. In the cloud context, multi-cloud architecture is an alternative solution, by which different cloud providers can be applied.

9.3.3   Access Control-Based Solutions

Using access control-based solutions is another popular method for securing cloud systems. This approach requires identification of all possible layers of access controls, such as access to servers, services, or databases. The mechanism of access control and entrance restrictions are the same. Only users who have the exact permitted authentication can access the resource. For service providers, they need to manage user authentications and design/organize access control processes, when access control-based solutions are applied.

One broadly accepted technique for improving access controls is using isolation techniques. This technique uses a logical container for implementing mobile apps or accessing computing resources. Logical containers are independent of each other. In the cloud context, *Virtual Machines* (VMs) are isolatable containers that ensure the computing resources in containers cannot be shared. The settings of VMs need to follow the two criteria, which are *Quality-of-Service* (QoS) and enforcement policies.

The next common approach of access controls for mobile apps is called *Customer-Managed Access Control* (CMAC). This method allows users to keep partial rights to configure the authentications of the access control. Data users need to match the requirement/criterion created by the customer to access data. The control parts are usually achieved by configuring settings of the customer account. For example, patient health records systems and Facebook offer the functions of Security & Privacy settings, which enable cloud customers to determine who can have access to their data. However, this paradigm requires a solid trustable relationship between cloud customers and service providers. On the cloud side, cloud customers' data are still reachable for service providers in many situations, even though the authentications are configured.

Another method of access control is using identification management to assist cloud customers to manage multiple accounts attached to multiple service providers. The operating principle of this approach is mapping cloud services by using business process modeling techniques. Based on the awareness of the business processes, cloud customers can determine what data can be shared with each service provider. Those unnecessary identities that are not required by the service offerings will be removed.

Figure 9.6 and Figure 9.7 illustrate a basic example of identification management, which shows the fundamental method of reducing identities. The implementation scenario is assumed at the service process of purchasing an item on Amazon. A few participants involved in the business process include Amazon, a credit card company, an arranger, and the UPS delivery. Figure 9.6 represents the stage of the identification management before identities are removed. Each participant collects mobile cloud customers' information by accessing their accounts. The collected data may increase the chance of privacy leakage since attacks can take place at each involved party.

Figure 9.6 Example of identification management: before removing identities.

Figure 9.7 represents the improvement of using identification management after identities are removed. Data users in the business process only have limited access to data. The remainder of the data access follows the business policies, which ensures that the service can be successfully delivered. It requires a deep understanding of the business process by identifying all sub-processes, such that the access of each party is configured to a limited scope. The chance of privacy leakage will be reduced since each party holds limited data access.

Figure 9.7 Example of identification management: after removing identities.

9.4 FURTHER READING

1. Y. Li, K. Gai, Z. Ming, H. Zhao, and M. Qiu. Intercrossed access controls for secure financial services on multimedia big data in cloud systems, *ACM Transactions on Multimedia Computing, Communications, and Applications (TOMM)*, *12*(4s), 67, 2016.

2. K. Gai, M. Qiu, L. Tao, and Y. Zhu. Intrusion detection techniques for mobile cloud computing in heterogeneous 5G. *Security and Communication Networks*, vol. 9, no. 16, pp. 3049–3058, Wiley, 2016.

3. K. Gai, M. Qiu, H. Zhao and J. Xiong. Privacy-aware adaptive data encryption strategy of big data in cloud computing. In *The IEEE 3rd International Conference on Cyber Security and Cloud Computing (CSCloud)*, pages 273–278, Beijing, China, 2016.

4. K. Gai. L. Qiu, H. Zhao, and M. Qiu, SA-EAST: Security-aware efficient data sharing and transferring scheme for intelligent transportation systems in mobile heterogeneous cloud computing. *ACM Transactions on Embedded Systems (TECS)*, vol. 16, no. 2, pp. 60, ACM, 2016.

9.5 SUMMARY

In this chapter, we learned a number of crucial aspects of protecting privacy in mobile clouds. When we talk about the improvements of the privacy protections, we need to understand the crucial security dimensions concerning protection techniques. Four dimensions have been addressed in this chapter, including infrastructure security, mobile data security and storage, mobile identity and access management, and privacy protection concerns. Based on these dimensions, we have reviewed a few common solutions to security and privacy issues in mobile clouds. Two main solutions include monitoring-based and access control-based solutions.

9.6 EXERCISES

1. What are three layers of infrastructure security in mobile cloud computing?

2. Describe when it is a challenge when an encryption method is used for protecting DiT from data leakage in some situations.

3. What are differences between *Data-in-Transit*, *Data-at-Rest*, and *Data-in-Use*?

4. Briefly describe what the security of the host layer concerns.

5. What is a *Data Lineage*?

6. What is a *Data Provenance*? Give an example.

7. What are the differences between *Managed Cloud Storage* and *Unmanaged Cloud Storage*?

8. Briefly describe the concept of the data life cycle.

9. Give three crucial concern dimensions in privacy issues and explain why they are important.

10. Describe the operating principle of *Monitoring-Based Solutions*.

11. Describe the mechanism of access control-based solutions.

12. Picture yourself as a mobile cloud customer who intends to purchase a product from an e-commerce website. The online trading

(e-commerce) company is called "E-COM." Your personal information includes name, birthday, phone number, e-mail, password, billing address, shipping address, and credit card. You need to buy an item called "ITEM" that is provided by the company "XYZ." The credit card service available at E-COM is provided by the "CRED" company. The physical item arranger organization is called "ARRAN." The deliver company is called "DELIV."

Your mission:

In this case, try to use identification management and create two diagrams. One diagram shows the business process as well as the corresponding data before removing identities. The other diagram shows the required data for each participant after removing identities. Please refer to the original status of data access for each party in Table 9.1

Table 9.1 Mapping Table of Party and Data Access for Exercise 12.

Party	Data Access
E-COM	name, birthday, phone number, e-mail, password, billing address, shipping address, and credit card
CRED	Name, birthday, billing address, credit card
ARRAN	Name, birthday, e-mail, shipping address
DELIV	Name, e-mail, phone number, shipping address
XYZ	Name, e-mail, phone number, shipping address

9.7 GLOSSARY

Data-at-Rest means the data state when the data are stored/recorded on the cloud-based physical server/media in mobile cloud computing.

Data-in-Transit means all data flowing over both trusted and untrusted networks, which usually take places between two nodes on the network.

Data-in-Use means the data state when the data are attached to one node in a network rather than the DaR state. Some examples of the particular nodes in a network include processor cache, disk cache, and memory.

Data Lineage refers to an approach defining a data life cycle that shows when and where the data are located.

Data Provenance is a term describing computational accuracy when data integrity is used.

Local Host refers to mobile devices used by mobile end users.

Managed Cloud Storage is a type of data storage deployment that provisions raw disk services either on-site or off-site so that users can have authorizations to configure the disk, such as partitions and formats.

Network Layer (in Infrastructure Security) refers to the security and privacy issues that exist or occur during the data transmissions on wireless networks.

Unmanaged Cloud Storage is a type of data storage deployment that provisions preconfigured storage so that users do not have authorizations to configure the disk, such as formatting or installing file systems and changing drive properties.

Wi-Fi Protected Setup is an emerging security standard for establishing a secure wireless home network.

WPA2-Enterprise is an enterprise-oriented security protocol that requires usernames and the corresponding passwords, such that one user can have multiple accounts.

WPA2-Personal is a security protocol that is designed for home users or small organizations, which only uses a single password.

IV

Integrating Service-Oriented Architecture with Mobile Cloud Computing

Web Services in Cloud Computing

CONTENTS

Web SERVICES is fundamental for delivering cloud computing services. The device-to-device services also empower the implementations of *Service-Oriented Architecture* (SOA). In this chapter, we will teach the relationships among cloud computing, SOA, and Web services, as well as explain the significance of integrating SOA with mobile cloud computing. A number of Web service specifications will be introduced in this chapter, too. After reading this chapter, students will be able to answer the following questions:

1. Why do we integrate *Service-Oriented Architecture* (SOA) with cloud computing?

2. What are the differences between SOAP and REST?

3. What is WS-Coordination?

4. What is an SOA? What does service mean in SOA?

5. What is the method of integrating SOA with cloud computing?

6. What is the BPEL4WS?

7. What is the WS-security specification? How do we form a WS-security framework?

10.1 INTRODUCTION

This chapter mainly covers two aspects. First, we have an overview of *Service-Oriented Architecture* (SOA) in Section 10.2. A group of concepts are introduced, including Web services [93, 94, 95, 96, 97, 98, 99, 71], We service specifications, and concepts related to SOA [3, 18, 100]. The differences between Web service specifications are clarified via comparisons.

The second aspect is the method of integrating SOA with cloud computing presented in Section 10.3. We introduce the fundamental information about integration. Then, two dimensions are addressed, namely, the business process execution language for Web services and Web service security specifications.

10.2   OVERVIEW OF SERVICE-ORIENTED ARCHITECTURE

10.2.1   Web Services

10.2.1.1   Basic Concepts

The concept of *Web Services* (WS) is a non-proprietary service offering between devices through the *World Wide Web* (WWW). The implementations of WS can support *Service-Oriented Architecture* (SOA) by using *Extensible Markup Language* (XML) and Internet protocols. Using WS can build machine-to-machine communications, which are convenient for transforming legacy applications into service providers.

Moreover, WS support system integrations are based on heterogeneous platforms and languages. In general, WS is considered an approach running on the host server, which is designed for client programs to invoke. The publicized method's signature is described by a *Web Service Definition Language* (WSDL) file that is automatically created by a tool. A client obtains the web services WSDL file either from a *Universal Description, Discovery, and Integration* (UDDI) registry or directly from the Web. For completing this process, a tool transforms the WSDL file into a proxy class, by which it uses the same method signature as that on the server.

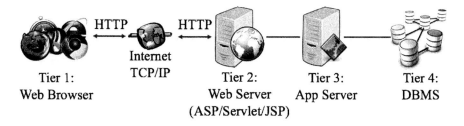

Figure 10.1   Four-tiered web architecture.

10.2.1.2   Web Services Architecture

Figure 10.1 represents the four-tiered Web architecture. As illustrated in the figure, the four tiers include Web browser, Web server, app server, and *Database Management System* (DBMS). The *Web Browser* tier is a service representation form/interface. In mobile cloud computing, this tier is generally operated on the mobile end. The tier Web browser and the tier Web server are connected via the Internet *Transmission Control Protocol/Internet Protocol* (TCP/IP) and the *Hyper-*

text Transfer Protocol (HTTP). A *Web Server* is the computing equipment with an IP address, which delivers Web pages and exclusively handles HTTP requests/protocols. Most web servers have domain names. Next, an *App Server* is a software framework offering both establishment and operating environments to web applications, which provides a complete service layer model. Finally, DBMS is a software/system that is designed, developed, and used for managing databases. In the context of cloud computing, DBMS is usually located at the physical machine layer. The architecture provides an essential structure for building a machine-to-machine service mechanism. Box 10.2.1.2 shows the methods of the Web service requirement analysis.

Box 10.2.1.2: Web Service Requirement Analysis Methods

> Currently, there are a variety of methods available for analyzing WS requirements in software systems. Two primary methods of gathering requirements are [101]:
>
> *Gathering Usage Scenarios* (GUS) analysis method: This refers to a group of methods that are designed to collect, analyze, and examine the application implementations for achieving a certain goal, which is usually to connect business and user requirements.
>
> *Critical Success Factor* (CSF) analysis method: This is a managerial approach that assists enterprise/organizations in making decisions by determining the necessary elements for successful achievements/goals.

For further understanding of the mechanism of Web services, Figure 10.2 shows the service process of Web services, which demonstrates the

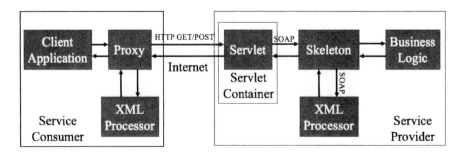

Figure 10.2 Service process of Web services.

main required tasks during the service delivery process between service consumers and service providers. In line with the figure, a *Simple Object Access Protocol* (SOAP) message is created by the proxy method body once the client calls the proxy's method. Simultaneously, the client call uses an HTTP POST request to send the SOAP file to a Web component, such as a servlet or *Active Server Pages* (ASP).

Next, the Web component takes responsibility for transforming the SOAP message into a method call that can be used for the local Web service implementation methods. The method is returned to the proxy object as a part of the HTTP response. The proxy method body parses the SOAP response message in terms of its own language before it returns the message to the client.

The following section reviews three WS specifications.

10.2.2 Specifications of Web Services

WS Specifications refer to a group of WS framework that are used. We have a quick review of three basic specifications for WS in this section: SOAP, *Representational State Transfer* (REST), and *JavaScript Object Notation* (JSON). Two advanced specifications are not briefly introduced in this section, including WS-Coordination and WS-Transaction.

10.2.2.1 Simple Object Access Protocol

In this section, we introduce a common WS protocol, SOAP, as well as a few basic related protocols, including WSDL and UDDI. SOAP, a type of WS specification, is a messaging protocol that uses HTTP and XML to enable programs to be operated on distinct operating systems. For example, using SOAP can support an app to run on Windows, Linux, and MAC OS 10. SOAP messages are generally considered hard-coded messages, which can be created without using a repository. Figure 10.3 represents the operating principle of SOAP. The service consumer is the party that sends out the SOAP service request. The service provider is the party that sends back the XML service response.

In addition, SOAP is one of the most popular messaging protocols used in XML-based WS. It establishes a standard message format. SOAP messaging is formed by XML documents that support the storage of *Remote Procedure Call* (RPC) or Document-Centric Data (DCD). Both synchronous and asynchronous data exchange models can be operated using SOAP.

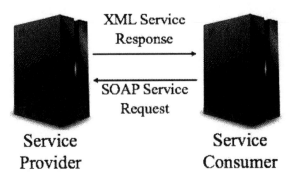

Figure 10.3 SOAP operating principle.

Figure 10.4 illustrates the basic structural organization of SOAP envelope, header, and body. In line with the figure, we provide the concept of each component in Box 10.2.2.1.

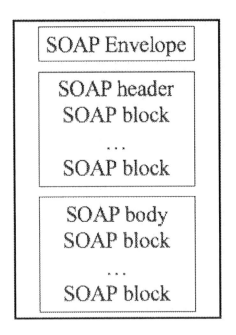

Figure 10.4 Illustration of SOAP envelope, header, and body.

Box 10.2.2.1: SOAP Envelope, SOAP Header, and SOAP Body.

SOAP Envelope is an element of the SOAP that defines the XML document by describing its boundaries.

SOAP Header is an optional element of SOAP that contains application-specific information, by which puts directives to SOAP processors and receives messages. A legitimate SOAP Header should be the first child element of the SOAP Envelope.

SOAP Body is a SOAP element that stores actual data or message.

Figure 10.5 represents an example of SOAP protocol binding, which shows the HTTP header information.

SOAPAction = "urn:soaphttpclient-action-uri"

Host = localhost

Content-type = text/xml; charset = utf-8

Content-length = 701

Figure 10.5 Illustration of SOAP protocol binding.

Furthermore, WSDL is specified by W3C and the definition is "an XML format for describing network services as a set of endpoints operating on messages containing either document-oriented or procedure-oriented information" [102]. WSDL provides the crucial language of describing WS, which ensures the consistency of WS such that WS can be found by and interfaced with other WS/applications.

UDDI is an XML protocol represented as an XML-based registry that is used for make WS discoverable on the Internet and integrable with other WS applications by providing register mechanisms. Making WS discoverable means that other WS, the service requestors, are able to find the WS, as well as their descriptions. The service requestors can be either the active requestors or the potential requestors. UDDI specification is an abstract of establishing a WS framework for hosting service descriptions by a centralized directory/registry. The critical part of UDDI is using the standardized profile records in order to satisfy different WS implementations, such as storing in a directory. Some examples of entities involved in the UDDI descriptions are business entities, specification pointers, service types, business relationships, and subscriptions.

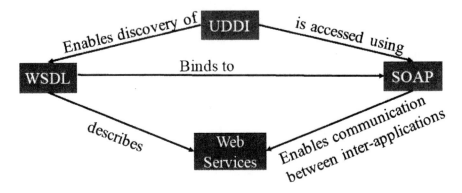

Figure 10.6 Relationships among UDDI, WSDL, SOAP, and Web services.

Figure 10.6 represents the evolutional relationships among SOAP, UDDI, WSDL, and WS. SOAP plays a crucial role in enabling WS communications between inter-applications.

10.2.2.2 Representational State Transfer

REST is a type of software architecture style that appeals to developers for developing WS in the WWW. Essentially, REST is a simpler type than SOAP. Considering the operations, REST is also easier to use than SOAP. These features have driven REST to be a popular *Application Programming Interface* (API) in contemporary cloud solutions, such as Amazon, Microsoft, and Google. The implementation of REST is also mobility friendly, which empowers the efficiency of communications by limiting verbosity.

However, the drawback of using REST is that it is difficult for creating a client from the metadata created on server side. Another form of WS, SOAP, can overcome this by using WSDL. There are a few differences with benefits and drawbacks between SOAP and REST. A comparison is given in Box 10.2.2.2. Figure 10.7 represents the operating principle of REST. According to the figure, service consumers send out REST service requests to the service provider. The service provider sends an XML service response back to the service consumer.

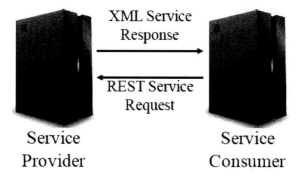

Figure 10.7 REST operating principle.

SOAP vs. REST

SOAP and REST perform differently in many aspects. We present a comparison between SOAP and REST here:
SOAP can do a better job in

1. independent language, platform, and transfer, since REST needs using HTTP;

2. distributed operating environments, since REST mainly supports point-to-point communications;

3. standardized messaging format;

4. providing pre-built extensibility;

5. handling built-in errors; and

6. performing automations while certain language products are applied;

REST can do a better job in

1. interacting with WS without the need of expensive tools;

2. being learning-friendly;

3. achieving efficient performance by using smaller-sized message formats without any extensive processing; SOAP applies XML on all messages;

4. having a similar design philosophy to other Web technologies.

10.2.2.3   JavaScript Object Notation

JSON is a type of lightweight data-interchange format that uses a subset of JavaScript rather than using XML. Name/value pairs are applied in JSON. The main benefit of using JSON is that it is easy for both human read and write actions and for machine parsing and generating operations. Figure 10.8 represents the operating principle of JSON. The service consumers send out a JSON service request to the service providers. The service providers send a JSON service response back to the service consumers.

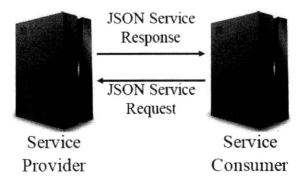

Figure 10.8   JSON operating principle.

10.2.2.4   WS-Coordination

WS-Coordination is a WS specification that provides a framework of context management in which the activity's coordination context can be tracked for the purpose of supporting distributed applications. The actions of the distributed applications are coordinated by the protocols that are provided by the extended framework in WS-Coordination. Moreover, WS-Transaction is a WS specification that is used to support WS-Coordination by describing coordination types. These two specifications are used to understand the integrity of the persistent activity context in distributed WS environments. Figure 10.9 represents the workflow of WS-Coordination. In the coordinator, the main tasks include creating context, registering for coordination, and selecting protocols.

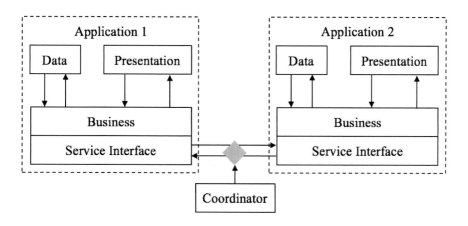

Figure 10.9 Workflow diagram of WS-Coordination.

10.2.3   Service-Oriented Architecture

10.2.3.1   Basic Concepts of SOA

Let's take a look at the relationships among *Service-Oriented Architecture* (SOA), cloud computing, and WS before we talk about SOA's concept and mechanism. Figure 10.10 represents the brief relationships of SOA, cloud computing, and WS by showing the overlaps.

As illustrated in the figure, SOA and WS are two technical domains that have a big intercrossed scope. Implementing cloud computing can be considered a type of WS application. Meanwhile, WS can be integrated with SOA, such that some cloud-based applications can also be considered SOA-based solutions. Some SOAs apply WS techniques that empower the cloud computing application; thus, this type of service offering SOA with WS integrates cloud computing services. Being aware of the roles of each entity in the loop can aid students to have a holistic picture of the concepts. The following statements introduce the definitions and mechanisms of SOA.

10.2.3.2   Understanding Services

SOA is an architectural style that is used to loosely couple interacting software agents. In the concept of SOA, a *Service* refers to a work unit that offers a well-defined self-contained function to realize a certain result desired by the service consumer, which needs to be accomplished

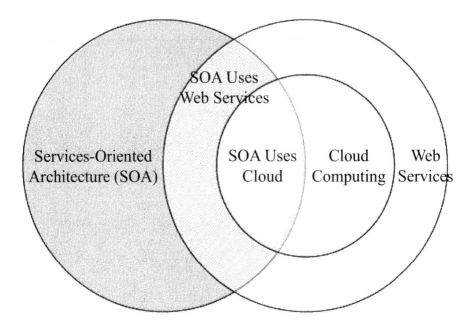

Figure 10.10 Relationships of SOA, cloud computing, and Web services.

by the service provider. A service needs its independent blocks and is able to notice other services, such that the application environment can be firmed. It implies that each service is responsible for a certain domain that is usually a specific business function or related functions. Generally speaking, both providers and consumers can be considered roles played by software agents on behalf of their owners.

Figure 10.11 represents the relations between the service provider and service consumer in SOA. The role of service providers and service consumers depend on the content of the message. A service provider sends out a service response to a service consumer. A service consumer sends out a service request to a service provider. Two common contexts exist in this relationship. First, both service consumers and providers understand the service requests and the subsequent responses that occur between them. Second, the roles of service providers and consumers are exchangeable.

In SOA, besides the service provider and service requester (consumer), there is another role, which is the service broker. A *Service Broker* is a service registry that stores the service information, including availability, usage, and state. Figure 10.12 represents a high-level

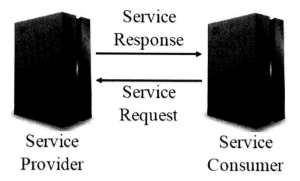

Figure 10.11 Service provider and service consumer in *Service-Oriented Architecture* (SOA).

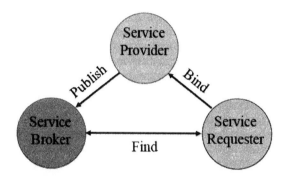

Figure 10.12 High-level *Service-Oriented Architecture* (SOA) showing service provider, service requester, and service broker.

diagram showing the interrelations among service providers, requesters, and brokers. Service providers publish the service with the service broker and enable service requesters to find the offered services. Service requesters bind to the service provider once the service is found at the service broker.

Figure 10.13 represents an example of the logical SOA reference architecture that derives from IBM. A common logical representation of SOA is given in Figure 10.14. Integrating cloud computing with SOA can enrich the benefits by sharing a few characteristics. The main advantages include:

1. **Service Visibility:** This refers to the difficulty level of finding

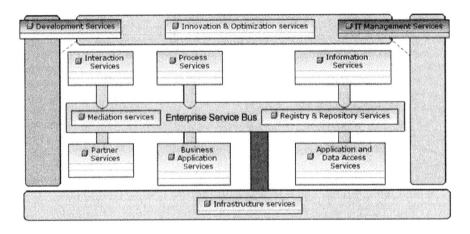

Figure 10.13 Logical SOA reference architecture from IBM. The figure is retrieved from https://www.ibm.com/.

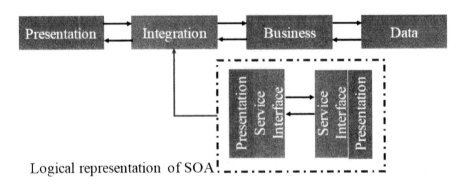

Figure 10.14 A common logical representation of SOA.

the available service. Integrating cloud computing with SOA can enable XXX-as-a-Service to reach a bigger group of service requesters by using service brokers in SOA.

2. **Service Governance:** Integrating cloud computing with SOA can empower both SOA design-time and runtime governance while XXX-as-a-Service is applied.

3. **Composability:** Using SOA can enable cloud-based solutions to compose legacy and updated computing resources.

4. **Functionality:** SOA can empower business automation solutions by supporting complex designs.

For showing the concept of integrating cloud computing with SOA, we give a structural mapping example in Figure 10.15.

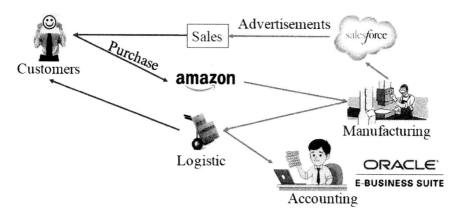

Figure 10.15   A simple example of integrating cloud computing and SOA.

10.3   INTEGRATING SERVICE-ORIENTED ARCHITECTURE WITH CLOUD COMPUTING

In this section, we introduce the basic mechanisms and techniques of integrating SOA with cloud computing.

10.3.1   Integration Fundamentals

For an enterprise/organization, a holistic value needs to be defined before migrating to the integrations of SOA and cloud computing. For example, determining a business dimension is a critical aspect, such as reducing operational costs, increasing reuse, and enhancing agility. Extending the organization to mobile cloud-based solutions can be executed after the value is positioned.

The integration of SOA and cloud computing starts with a series of decision-making procedures. The decision-making missions mainly focus on defining targets or entities. Basic steps for start-up include defining data, defining services, defining processes, defining governance, and defining the deployments. Among these, defining deployments refers to whether the system uses mobile cloud-based solutions or on-premises applications. After all entities are determined by the organization, the integration can be executed by the following approach.

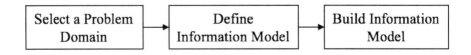

Figure 10.16   Basic steps of integrating cloud computing and SOA.

Figure 10.16 represents the basic steps of integrating cloud computing with SOA, which consists of three steps, including selecting a problem domain, defining an information model, and building the information model. The detailed descriptions of each step are given as follows:

1. **Selecting Problem Domains:** The purpose of this step is to extend the existing SOA to (mobile) cloud computing platforms. The value dimensions are determined in the start-up period, such that the goal is firmed at this step. Constraints need to be clarified by defining the reasonable number of systems or total execution time.

2. **Defining the Information Model:** At this step, we need to be aware of the complexity and entities of the problem. Understanding data is the basis of defining the information model, such as the data dictionary and meta data. For some complex problems, ontologies need to be understood as well. Finally, a data catalog needs to be completed.

3. **Building the Information Model:** An information model needs to be built based on the outcomes of defining the information models. There are two kinds of information models, which are the *Logical Model* and the *Physical Model.*

   • The *Logical Model* is a relationship diagram showing all entities and their interrelations.

   • The *Physical Model* is a practical approach examining the model, which is generally used for homogeneous database approaches.

The following section introduces the business process execution language for WS.

10.3.2 Web Services Business Process Execution Language

Business Process Execution Language for Web Services (BPEL4WS) is a substitute for *Enterprise Application Integration* (EAI). BPEL4WS is an executable language that specifies business activities/tasks by using WS. In BPEL4WS, WS-Coordination is used to offer a context-management framework; WS-Transaction is used to define activity coordination types and establish the coordination methods for managing long-running activities. Figure 10.17 represents the workflow diagram of BPEL4WS processes. The integration is operated by modeling business processes.

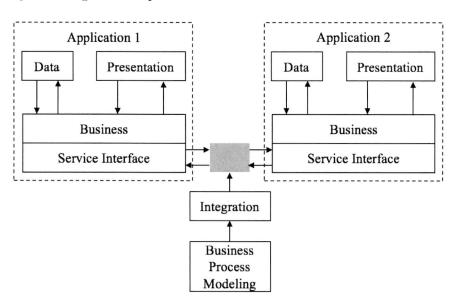

Figure 10.17 Workflow diagram for BPEL4WS process services.

10.3.3 Web Services Security and Specifications

10.3.3.1 Specifications

The security issue arises in WS because of the continuously growing complexity of the business logic and the implementation of risks. For example, integrating with an external business partner will dramatically increase the complexity, in which security problems possibly occur. General terms of security are introduced in Chapter 8. In this section, we focus on the security specifications for WS.

1. **XML Key Management Specification:**

   XML Key Management Specification (XKMS) is a Web specification that provides a standardized approach for gaining and managing public keys, by which *Public Key Infrastructure* (PKI) technologies are not required. Two standards are involved in XKMS, including the *XML Key Registration Service* and the *XML Key Information Service*.

2. **XML-Encryption and XML-Signature**

   XML-Encryption (XML-Enc) is a specification that provides a standardized model for encryptions. The target data include both binary and textual data. This is a method of information exchange for decryptions. Moreover, *XML-Signature* (XML-Sig) is an approach of standardizing the format for representing digital signature data.

3. **Secure Socket Layer**

   Secure Socket Layer (SSL) is a security technology that is used to secure transport-level security. SSL can protect message contents when the messages are transmitted between intermediate services. However, SSL cannot protect message contents when the execution is processed by an intermediate service.

10.3.3.2   WS-Security Framework

Box 10.3.3.1: Security Token

Security Token: It is a security mechanism that connects service interfaces between applications by providing a set of claims for communicating security statuses. A claim is an assertion of the service requester that is validated by service providers. Figure 10.19 represents an example of applying security token in WS-security framework by using SOAP messages.

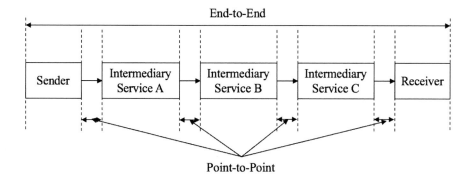

Figure 10.18 End-to-end and point-to-point security solutions.

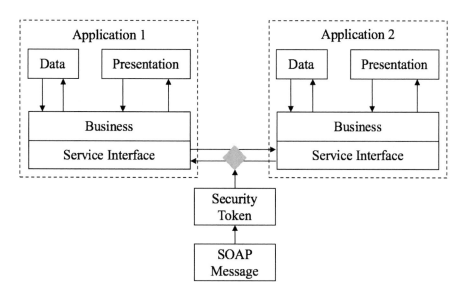

Figure 10.19 An example of applying security token in WS-security framework by using SOAP messages.

A WS-security framework is an establishment of supplementary security-oriented specifications by fundamental security standards. The primary purpose of using a WS-security framework is interconnecting distinct security models. Generally speaking, a WS-security framework is designed to achieve an end-to-end security solution. An important technical part in the WS-security framework is the *Security Token* and its definition is given in Box 10.3.3.2. Figure 10.18 represents the difference between the end-to-end and point-to-point security solutions.

10.4 SUMMARY

In this chapter, we have discussed the implementations of WS in cloud computing from the perspective of SOA. We also defined the concepts of WS and SOA and presented both general WS specifications and specific WS specifications. Moreover, BPEL4WS was introduced, which was the approach for integrating SOA with cloud computing by modeling business processes. Finally, we presented a few WS-security specifications and methods of forming a WS-security framework.

10.5 EXERCISES

1. What is the concept of *Web Services*?

2. What is the concept of *Service-Oriented Architecture* (SOA)? Why can Web services support SOA?

3. What is the four-tiered Web architecture? Which four layers are in the architecture? How do they work?

4. What is a *Web server*?

5. Briefly explain the service process of Web services.

6. Give two methods of Web service requirement analysis with detailed statements.

7. Describe the concepts of SOAP, REST, and JSON in detail including their differences.

8. How do *SOAP Envelope*, *SOAP Header*, and *SOAP Body* work?

9. What is the concept of WSDL?

10. Briefly describe the relationships among UDDI, WSDL, SOAP, and Web services.

11. Explain the mechanism of WS-coordination.

12. What are the relationships of SOA, cloud computing, and Web services?

10.6 GLOSSARY

App Server is a software framework offering both establishment and operating environments to web applications, which provides a complete service layer model.

Business Process Execution Language for Web Services (BPEL4WS) is an executable language that specifies business activities/tasks using WS.

Critical Success Factor Analysis Method is a managerial approach that assists enterprises/organizations in making decisions by determining the necessary elements for successful achievements/goals.

Gathering Usage Scenarios Analysis Method refers to a group of methods that are designed to collect, analyze, and examine the application implementations for achieving a certain goal, which is usually to connect business and user requirements.

Logical Model is a relationship diagram showing all entities and their interrelation (in the information model).

Physical Model is a practical approach examining the model, which is generally used for homogeneous database approaches (in the information model).

Representational State Transfer is a type of software architecture style that appeals to developers for developing WS in WWW.

Security Token is a security mechanism that connects service interfaces between applications by providing a set of claims for communicating security statuses (in the WS-security framework).

Service refers to a work unit that offers a well-defined self-contained function to realize a certain result desired by the service consumer, which needs to be accomplished by the service provider (in SOA).

Service Broker is a service registry that stores the service information, including availability, usage, and state (in SOA).

Service-Oriented Architecture is an architectural style that is used to loosely couple among interacting software agents.

Simple Object Access Protocol a type of WS specification, is a messaging protocol that uses HTTP and XML to enable programs to be operated on distinct operating systems.

SOAP Body is a SOAP element that stores actual data or a message.

SOAP Envelope is an element of SOAP that defines the XML document by describing its boundaries.

SOAP Header is an optional element of SOAP that contains application-specific information, by which puts directives to SOAP processors and receives messages. A legitimate SOAP Header should be the first child element of the SOAP Envelope.

Universal Description, Discovery, and Integration is an XML protocol represented as an XML-based registry that is used to make WS discoverable on the Internet and integrable with other WS applications by providing register mechanisms.

WS-Coordination is a WS specification that provides a framework of context management in which the activity's coordination context can be tracked for the purpose of supporting distributed applications.

Web Server is the computing equipment with an IP address, which delivers Web pages and exclusively handles HTTP requests/protocols.

Web Service Definition Language is "an XML format for describing network services as a set of endpoints operating on messages containing either document-oriented or procedure-oriented information" (W3C definition).

Web Services is a non-proprietary service offering between devices through the *World Wide Web* (WWW).

WS-Transaction is a WS specification that is used to support WS-Coordination by describing coordination types.

V

Appendices

Sample of A Course Project

CONTENTS

A.1 PROJECT DESCRIPTION

Each student needs to propose a course project by *Month/Date/Year*, and complete the project with a presentation and written report at the end of the semester. A typical course project studies a particular sub-topic of the course, such as optimization of cloud computing, cloud computing security, or big data in cloud computing.

This course project consists of two parts, including hands-on experiences and a research essay. Please see the following instructions regarding the course project.

A.2 PART I: HANDS-ON EXPERIENCE

The purpose of this project is to provide students with an opportunity of to operate a typical cloud platform in an experimental setting. The detailed operation guide manual will be provided in a separate document. Please see the following experimental requirements to prepare your experiment.

A.2.1 Experimental Requirement

- **Phase 1:**

  1. Apply for an account for Amazon Web Services (Amazon EC2: https://aws.amazon.com/ec2/).

  2. Create a Windows 2012 Virtual Machine (VM) on Amazon EC2.

  3. Install a JRE on the VM you installed on Amazon EC2.1. (available at http://www.oracle.com/technetwork/java/javase/downloads/jre8-downloads-2133155.html).

  4. On your VM, run the program to simulate the task scheduling according to the DAG given in the experiment document. After you run the program, you can obtain the optimal solutions to scheduling tasks. Your mission is to record the result and the program execution time.

  5. On your device, repeat the 4th step and record the program execution time, too. Compare this execution time with the execution time needed by Step 4. Deadline: *Month/Date/Year*.

- **Phase 2:**

  1. Design a DAG and repeat Step 4 and 5 in Phase 1.

  2. Write a one-page report that clearly states your experimental configuration, experiment process, results, DAG design, and experimental analysis. Submit your report to the Course Work on Discussion Board.

A.2.2 DAG and Execution Time Table

This section provides students with a DAG as well as the execution time table for the first phase of the project experiment. Please see the following graphs and table for detailed information.

Note: The tool used in this project does not consider preemptable tasks. Every task has an equal priority.

The number of tasks: 12 (Tasks 0 – 11) The number of cloud servers: 3 (Cloud 0, Cloud 1, Cloud 3) Figure A.1: DAG used for project Phase I. Table A.1: Execution time table for Figure A.1.

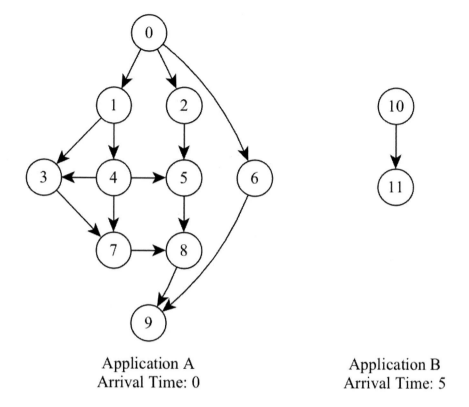

Application A
Arrival Time: 0

Application B
Arrival Time: 5

Figure A.1 DAG used for the project.

Table A.1 Execution Time Table.

Task	0	1	2	3	4	5	6	7	8	9	10	11
Cloud 0	3	5	7	10	6	4	3	8	6	13	12	2
Cloud 1	2	3	4	9	7	5	2	4	9	8	10	1
Cloud 2	5	4	5	12	8	9	4	5	12	5	11	3

A.3   PART II: RESEARCH ESSAY

This project is designed to assist students to deepen their understanding of cloud computing in a specific focus.

Project Requirement:

1. Students should finish a short paper at least two pages using

IEEE conference proceedings format. The length of your paper includes all text, figures, and references. The maximum length of the paper is six pages. The format template can be retrieved from the link: https://www.ieee.org/conferences_events /conferences/publishing/templates.html.

2. Your submission must represent original and unpublished work.

3. Students need to determine a research topic that is related to the content of this class. Cloud computing must be a focus in your short paper. A few topic samples are given for your reference as follows:

   (a) A review of crucial security issues in cloud computing and the main solutions.

   (b) Task scheduling optimizations for heterogeneous cloud computing.

   (c) Empirical study of cloud service model (IaaS, PaaS, or SaaS) implementations in industry.

   (d) Value creations of using cloud computing for enterprises in XXX industry.

   (e) Privacy issues and main solutions in cloud computing.

   (f) An analysis of cyber threats in cloud computing.

   (g) A survey of security policy and legal considerations in cloud computing.

   (h) Optimizations of the virtualization techniques in cloud computing.

4. In your paper, you should cover/answer the following questions:

   (a) What is your research question? Why is your question researchable?

   (b) What is the research background?

   (c) What is (are) the main contribution(s) of your work?

   (d) What are the recent research achievements related to your research question? (discuss related work:)

   (e) What is your solution? How do you evaluate your solution? What are the evaluation results?

A.4 TUTORIAL: CREATE/LAUNCH VIRTUAL MACHINE FOR PART I

This is a tutorial that teaches the methods of creating and launching the virtual machine for Part I.

1. Go to Amazon EC2: https://aws.amazon.com/ec2/.

2. You will see following page: Create your own account by clicking on **Create an AWS Account**. The registration instructions are available from: https://aws.amazon.com/documentation/gettingstarted/. Refer to Figure A.2.

Figure A.2 Screen capture for Step 2.

3. You can use your account to log in to the system if you already have one. Your can refer to Figure A.3 for this step.

4. After you log in to the system, you will see a list of services shown on the page. Click on EC2 at the top left of the page, as shown in Figure A.4.

5. Click on Launch Instance to select instance, which is shown in Figure A.5.

6. A list of virtual machines will be shown after you click on Launch

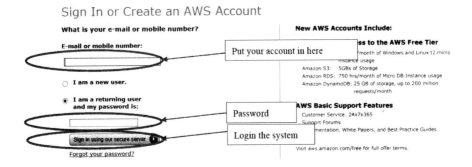

Figure A.3 Screen capture for Step 3.

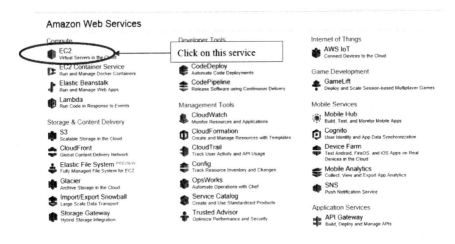

Figure A.4 Screen capture for Step 4.

Instance. Scroll down the page and select Microsoft Windows Server 2012 R2 Base. The screen capture is shown in Figure A.6.

7. You will see "Step 2: Choose an Instance Type." There are a number of virtual machine alternatives in the table. For this project,

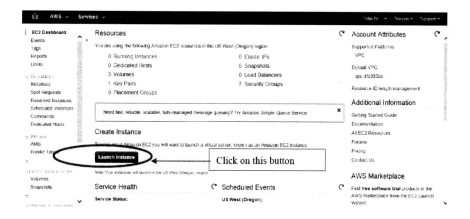

Figure A.5 Screen capture for Step 5.

Figure A.6 Screen capture for Step 6.

we only use its free service. Click on "Review and Launch." Please see the following screen capture in Figure A.7.

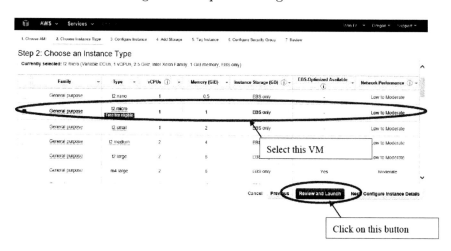

Figure A.7 Screen capture for Step 7.

8. You will see "Step 7: Review Instance Launch". Click on Launch button at the right bottom. Refer to Figure A.8.

Figure A.8   Screen capture for Step 8.

9. You will see the following dialog: Select "Create a new key pair." Create your own key pair name in the box under "Key pair name." Click on "Download Key Pair" and save the key pair. The screen capture is shown in Figure A.9.

10. Click on "Launch Instances" after you save your key pair. You can refer to Figure A.10.

11. You will see the screen as shown in Figure A.11. It provides you with your instance status. You can click on "View Instances."

12. You will see the instance you just launched. You select the VM and then click on "Connect." Refer to Figure A.12.

13. Please complete the following steps, which are shown in Figures A.13, A.14, and A.15.

14. Click on "Connect". Your VM will be launched after you click on OK. Refer to Figures A.16 and A.17.

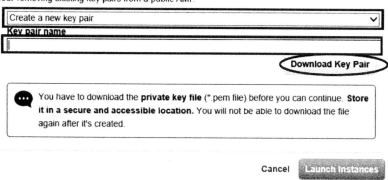

Figure A.9   Screen capture for Step 9.

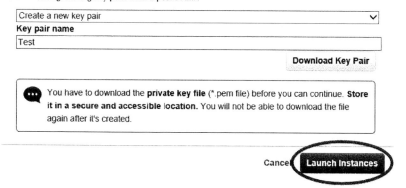

Figure A.10   Screen capture for Step 10.

Launch Status

> ✓ Your instances are now launching
> The following instance launches have been initiated: i-4289389a   View launch log

> ❶ Get notified of estimated charges
> Create billing alerts to get an email notification when estimated charges on your AWS bill exceed an amount you define (for example, if you exceed the free usage tier)

How to connect to your instances

Your instances are launching, and it may take a few minutes until they are in the **running** state, when they will be ready for you to use. Usage hours on your new instances will start immediately and continue to accrue until you stop or terminate your instances.

Click **View Instances** to monitor your instances' status. Once your instances are in the **running** state, you can **connect** to them from the instances screen. Find out how to connect to your instances.

▾ Here are some helpful resources to get you started
- Amazon EC2 User Guide
- How to connect to your Windows instance
- Amazon EC2 Microsoft Windows Guide
- Learn about AWS Free Usage Tier
- Amazon EC2 Discussion Forum

While your instances are launching you can also

Create status check alarms to be notified when these instances fail status checks. (Additional charges may apply)
Create and attach additional EBS volumes. (Additional charges may apply)
Manage security groups

View Instances

Figure A.11   Screen capture for Step 11.

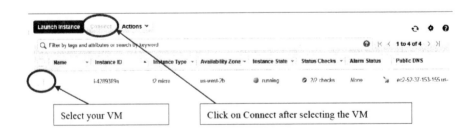

Figure A.12   Screen capture for Step 12.

Figure A.13   Screen capture for Step 13-1.

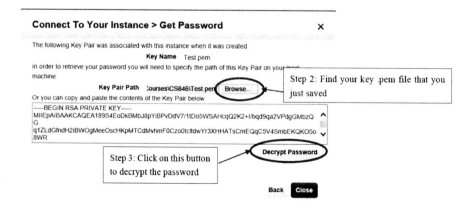

Figure A.14   Screen capture for Step 13-2.

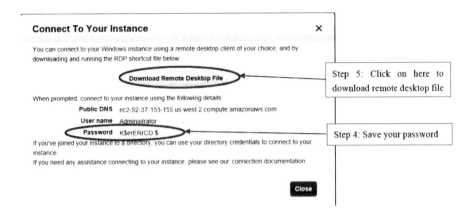

Figure A.15 Screen capture for Step 13-3.

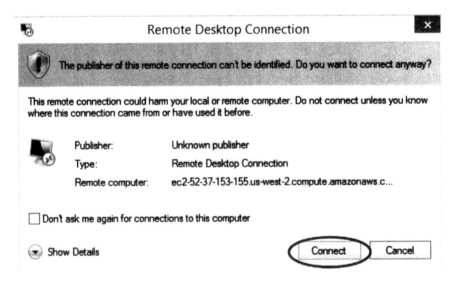

Figure A.16 Screen capture for Step 14-1.

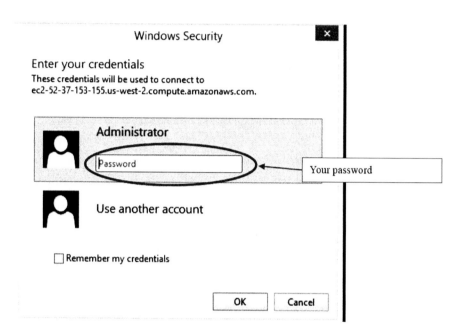

Figure A.17   Screen capture for Step 14-2.

Sample of Exam Sheet

CONTENTS

B.1   PART I. MULTIPLE CHOICE QUESTIONS

In this part, students need to select the correct answer for each question. The following is a sample question:

Among the following, which is NOT a key technology of cloud computing?

(A) Virtualization

(B) Parallel Programming

(C) Mass Distributed Storage

(D) Communications Protocol

You need to select "D" since D is the correct answer.

Questions:

- **Question 1** Among the following, which statement is INCORRECT?

  (A) Database-as-a-Service is an approach for enterprises to optimize database administration.

  (B) Database-as-a-Service can save database costs because of the scalable database usage.

  (C) Database-as-a-Service is absolutely secure in any situation.

  (D) Database-as-a-Service may have performance concern since sometimes virtualization cannot have the same capability or functionality as the local database.

- **Question 2** Among the following, which is an INCORRECT statement regarding Forwarding Workload approach?

  (A) The Forwarding Workload approach can extend/scale up the resource pool.

  (B) The Forwarding Workload approach supports operations done between private and public clouds.

  (C) The Forwarding Workload approach is usually designed to increase computation efficiency or reduce costs.

  (D) The Forwarding Workload approach is the ONLY optimization approach of Infrastructure-as-a-Service.

- **Question 3** Among the following, which is an INCORRECT statement about Big Data?

  (A) Big Data's 3V characteristics include volume, velocity, and variety.

  (B) Currently, Big Data is dealing with a computing efficiency issue that can be caused by various reasons, such as a large volume of data, great variety of data types, and distributed data sources.

  (C) Results of Big Data are always effective since we retrieve messy information from data mining.

  (D) Big Data is not only about data storage but also addresses data processing.

- **Question 4** Among the following, which is a CORRECT statement about the current security problem in cloud computing?

  (A) A solution to multi-tenancy issues in cloud computing is using cloud services from different cloud service providers.

  (B) Customers' loss of control can be caused by user identity management that is managed by cloud service providers.

  (C) Cloud customers rely on cloud service providers to ensure data security and privacy, resource availability, and to monitor or repair services or computing resources.

  (D) Lack of trust in cloud computing can be considered a third-party issue in most situations.

- **Question 5** Among the following, which is an INCORRECT statement?

  (A) The main components of a threat model include an attacker/adversary model, analysis of the attacker/adversary's goal, and examination of vulnerabilities/threats.

  (B) Insider attacks threaten customers' privacy because customers' behaviors and information can be sold to a third-party.

  (C) The security levels are always the same even though there are different cloud service models available.

  (D) Outsider threats can be done by various attack methods, such as probing network traffic and inserting malicious traffic.

- **Question 6** Among the following, which is a CORRECT statement about *Data Security* and *Storage*?

  (A) Crucial aspects of data security and storage include data generation, data transmission, and data processing.

  (B) Data lineage means knowing when and where the data are located, which is important for audit/compliance purposes.

  (C) Data provenance is about computation efficiency, which is important for data integrity.

  (D) Managed cloud storage is a type of preconfigured storage.

- **Question 7** Among the following, which is a CORRECT statement about Web Services?

  (A) Web Services are used for human-to-machine communications.

  (B) The signature of the publicized method is described by a Web Service Definition Language file, which is automatically generated by a tool.

  (C) A common four-tiered Web architecture consists of a user interface, Web server, App server, and physical machine.

  (D) Web Services support system integration based on homomorphous platforms and languages.

- **Question 8** Among the following, which is a CORRECT statement about Service-Oriented Architecture?

  (A) Service-Oriented Architecture is an architectural style whose goal is to achieve loose coupling among interacting software agents.

  (B) A Service is to achieve desired end results for a service provider.

  (C) Service Providers and Service Consumers are non-exchangeable.

  (D) *Business Process Execution Language for Web Services* (BPEL4WS) is an alternative for executing *Enterprise Application Integration* (EAI).

- **Question 9** Among the following which is an INCORRECT statement about integrating XML into applications?

  (A) Positioning XML data representation in the architecture can help us understand how the enterprise can receive benefits from both applications and integration architecture.

  (B) One of the advantages of self-descriptive XML documents is that they can be large.

  (C) For consistently applying XML, we should establish a standard data flow process.

  (D) Considering extensibility and reusability, we had better use generic element-type names, avoiding incorporating names for a company, because they may change.

- **Question 10** Among the following, which is a CORRECT statement about integrating Web Services into applications?

  (A) Utility Services is a service model that is used to encapsulate functionality, which can be reused within and between applications.

  (B) Business Services is a service model that is designed to represent generic business functions, which combine the existing functionality into general business contexts.

  (C) Controller Services is a service model that contains a specific logic to complete a certain representation of the business function.

  (D) An SOA always contains only one Business Service.

B.2   PART II. SHORT ANSWER QUESTION

- **Question 11** What are three specifications for Web Services mentioned in our class? [Hints: SOAP, REST, JSON]

- **Question 12** Try to give detailed descriptions of these three specifications and explain how these specifications influence hybrid cloud computing performance.

- **Question 13** What are differences between SOAP and REST?

- **Question 14** Answer the following questions about integrating SOA with cloud computing:

  1. What are the relations between Web Services and Cloud Computing? What are the relations between SOA and Cloud Computing?

  2. Explain why we need to integrate SOA with cloud computing.

  3. What are the basic steps of integrating SOA with cloud computing?

- **Question 15** Answer the following questions about integrating Web Services with Applications:

1. What are the main 10 processes of modeling Service-Oriented component classes? (Please list key words for each step.)

2. What are the main differences between Parameter-Driven and Operation-Oriented interfaces?

3. What are two types of granularity for Web Services interfaces mentioned in our class? Give some detailed descriptions.

- **Question 16** Please use the Cloud List Scheduling algorithm (Greedy algorithm) to calculate the total execution time of Applications 1, 2, and 3. Parallel computing needs to be considered. Detailed steps are required.
  Conditions:

  1. Application 1 is a Best-Effort application that has 5 tasks. Arrival time is 0.

  2. Applications 2 and 3 are two Advance Reservation applications that have 7 tasks in total. Application 2's arrival time is 3 and Application 3's arrival time is 9.

  3. There are three cloud servers, Clouds 1, 2, and 3.

  4. The given figure (Figure B.1) shows the *Directed Acyclic Graph* (DAG) for Applications 1, 2, and 3.

  5. Table B.1 gives the execution time table.

Table B.1 Execution Time Table.

	A	B	C	D	E	F	G	H	I	J	K	L
Cloud 1	2	6	5	7	5	4	8	2	4	8	9	2
Cloud 2	3	8	3	10	9	2	8	3	5	4	3	3
Cloud 3	5	4	8	5	2	3	4	6	7	6	7	4

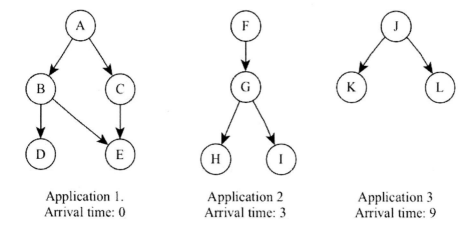

Application 1.
Arrival time: 0

Application 2
Arrival time: 3

Application 3
Arrival time: 9

Figure B.1 DAGs for Applications 1, 2, and 3.

B.3 ANSWERS FOR PART I

Table B.2 Answers for Part I of Appendix B.

Question	1	2	3	4	5	6	7	8	9	10
Answer	C	D	C	A	C	C	B	A	B	A

Simulator Tool Sample Codes

CONTENTS

This appendix provides sample Java codes [1] for the project tool used in the *Hands-On Experience* project that is given in Appendix A. The codes use a *Brute Force* algorithm that traverses all possibilities and obtain the optimal solution. Three programs include: UI.java, Task.java, and TaskGraph.java. Among the programs, UI.java is written for the user interface; Task.java is a class for recording all tasks' information; TaskGraph.java is written for recording the class of the entire DAG. Figure C.1 shows an example of the output result by running the program.

---

[1]This program was written by Professor Hui Zhao, Software School, Henan University, China.

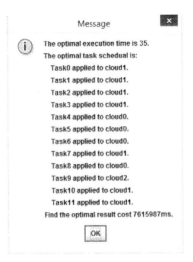

Figure C.1 An example of the output result.

C.1 PROGRAM 1: USER INTERFACE (UI.JAVA)

```java
import java.awt.Dimension;
import java.awt.FlowLayout;
import java.awt.event.ActionEvent;
import java.awt.event.ActionListener;
import java.util.ArrayList;
import java.util.List;

import javax.swing.JButton;
import javax.swing.JCheckBox;
import javax.swing.JFrame;
import javax.swing.JLabel;
import javax.swing.JOptionPane;
import javax.swing.JPanel;
import javax.swing.JScrollPane;
import javax.swing.JTextField;

public class UI extends JFrame {
    public static final int WIDTH = 600;
    public static final int HEIGHT = 400;

    private int count = 0;
    private TaskGraph tg;
```

```java
private int taskNum;
private int cloudNum;

private JTextField[] txtExeTimes;
private JCheckBox[] chkPriors;
private JTextField txtArrivalTime;
private JLabel lblTaskName;

public UI(int taskNum, int cloudNum) {
    this.taskNum = taskNum;
    this.cloudNum = cloudNum;
    this.tg = new TaskGraph(cloudNum);

    this.setSize(WIDTH, HEIGHT);
    this.setTitle("Task_detail");
    this.setDefaultCloseOperation(JFrame.
        DO_NOTHING_ON_CLOSE);
    FlowLayout flowLayout = new FlowLayout();
    flowLayout.setAlignment(FlowLayout.LEFT);
    this.setLayout(flowLayout);

    lblTaskName = new JLabel("Task" + this.count + "
        :"
            + "_____
                _____"
            + "_____
                _____"
            + "_____
                _____"
            + "_____
                _____");
    this.getContentPane().add(lblTaskName);

    JLabel lblCloudExeTime = new JLabel("Execution_
        time(s):");
    this.getContentPane().add(lblCloudExeTime);

    JPanel pnlExeTimeTable = new JPanel(new
        FlowLayout(FlowLayout.LEFT, 1,
                        1));
    JScrollPane sclExeTimeTable = new JScrollPane(
        pnlExeTimeTable);
    sclExeTimeTable.setPreferredSize(new Dimension(
        WIDTH - 25, HEIGHT / 4));
```

```
sclExeTimeTable
                .setHorizontalScrollBarPolicy(
                    JScrollPane.
                    HORIZONTAL_SCROLLBAR_NEVER);
sclExeTimeTable
                .setVerticalScrollBarPolicy(
                    JScrollPane.
                    VERTICAL_SCROLLBAR_AS_NEEDED
                    );

txtExeTimes = new JTextField[cloudNum];
int pnlExeTimeTableHeight = 55;
for (int i = 0; i < cloudNum; i++) {
        JLabel name = new JLabel("  Cloud" + i +
            " : ");
        txtExeTimes[i] = new JTextField(4);
        txtExeTimes[i].setText("0");
        pnlExeTimeTable.add(name);
        pnlExeTimeTable.add(txtExeTimes[i]);
        pnlExeTimeTableHeight += 5;
}
pnlExeTimeTable.setPreferredSize(new Dimension(
    WIDTH - 25, pnlExeTimeTableHeight));
this.add(sclExeTimeTable);

JLabel lblArrivalTime = new JLabel("Arrival time
    :");
this.getContentPane().add(lblArrivalTime);

txtArrivalTime = new JTextField(40);
txtArrivalTime.setText("After predecessor(s)
    finish");
this.getContentPane().add(txtArrivalTime);

JLabel lblPriors = new JLabel("Predecessor:");
this.getContentPane().add(lblPriors);

JPanel pnlPriors = new JPanel(new FlowLayout(
    FlowLayout.LEFT, 1, 1));
JScrollPane sclPriors = new JScrollPane(
    pnlPriors);
sclPriors.setPreferredSize(new Dimension(WIDTH -
    25, HEIGHT / 4));
sclPriors
```

```
                    .setHorizontalScrollBarPolicy(
                        JScrollPane.
                        HORIZONTAL_SCROLLBAR_NEVER);
    sclPriors
                    .setVerticalScrollBarPolicy(JScrollPane.
                        VERTICAL_SCROLLBAR_AS_NEEDED);

    chkPriors = new JCheckBox[taskNum];
    int panelHeight = 55;
    for (int i = 0; i < taskNum; i++) {
            chkPriors[i] = new JCheckBox("Task" + i
                + ",__");
            pnlPriors.add(chkPriors[i]);
            panelHeight += 5;
    }
    pnlPriors.setPreferredSize(new Dimension(WIDTH −
        25, panelHeight));
    this.add(sclPriors);

    JButton btnOK = new JButton("OK");
    btnOK.addActionListener(new ActionListener() {
            public void actionPerformed(ActionEvent
                e) {
                    count++;
                    if (count < UI.this.taskNum) {
                      UI.this.addTask(count − 1);
                      UI.this.reset();
                    } else {
                      UI.this.addTask(count − 1);
                      long time1 = System.
                          currentTimeMillis();
                      UI.this.tg.findOptimalSchedual
                          ();

                      long time2 = System.
                          currentTimeMillis();
                      UI.this.dispose();
                      StringBuilder msg = new
                          StringBuilder();
                      msg.append("The_optimal_
                          execution_time_is_"+ tg.
                          getMinExecutionTime() + ".\n
                          ");
                      msg.append("The_optimal_task_
                          schedual_is:\n");
```

```java
                                    int[] taskSchedual = tg.
                                        getMinAppliedServer();
                                    for (int i = 0; i < taskSchedual
                                        .length; i++) {
                                    msg.append("    Task" + i + " 
                                        applied to cloud "+
                                        taskSchedual[i] + ".\n");
                                    }

                                    msg.append("Find the optimal 
                                        result cost "+ (time2 -
                                        time1) + "ms.");
                                    JOptionPane.showMessageDialog(
                                        null, msg);
                                    }
                            }
                });

        this.add(btnOK);
        setSize(WIDTH, HEIGHT);
        setVisible(true);
    }

    private void addTask(int taskId) {
        int[] exeTime = new int[this.cloudNum];
        for (int i = 0; i < this.cloudNum; i++) {
                exeTime[i] = Integer.parseInt(this.
                    txtExeTimes[i].getText());
        }

        int canStartTime;
        try {
                canStartTime = Integer.parseInt(this.
                    txtArrivalTime.getText());
        } catch (NumberFormatException e) {
                canStartTime = Integer.MAX_VALUE;
        }
        List<Integer> prior = new ArrayList<Integer>();
        for (int i = 0; i < this.taskNum; i++) {
                if (this.chkPriors[i].isSelected()) {
                        prior.add(i);
                }
        }
        this.tg.add(new Task(taskId, exeTime,
            canStartTime, prior));
```

```java
}

    private void reset() {
        lblTaskName.setText("Task" + this.count + ":"
            + " ~~~~~~~~~~~~~~~~~~~~~~~~~~~~~~~~~~~~~~~~~~~
            "
            + " ~~~~~~~~~~~~~~~~~~~~~~~~~~~~~~~~~~~~~~~~~~~
            "
            + " ~~~~~~~~~~~~~~~~~~~~~~~~~~~~~~~~~~~~~~~~~~~
            "
            + " ~~~~~~~~~~~~~~~~~~~~~~~~~~~~~~~~~~~~~~~~~~~
            ");
        for (int i = 0; i < this.cloudNum; i++) {
            this.txtExeTimes[i].setText("0");
        }
        this.txtArrivalTime.setText("After_predecessor(s
            )_finish");
        for (int i = 0; i < this.taskNum; i++) {
            this.chkPriors[i].setSelected(false);
        }
    }
}

    public static void main(String[] args) {
        String strTaskNum = JOptionPane.showInputDialog(
                "How_many_tasks_do_you_have?", "
                1");
        int intTaskNum = Integer.parseInt(strTaskNum);
        String strCloudNum = JOptionPane.showInputDialog
            (
                "How_many_cloud_servers_do_you_
                have?", "1");
        int intCloudNum = Integer.parseInt(strCloudNum);
        new UI(intTaskNum, intCloudNum);
    }
}
```

C.2 PROGRAM 2: TASK.JAVA

```java
import java.util.List;

public class Task {
        int id;
        private int[] executionTime;
```

```java
private int canStartTime;
private int iniCanStartTime;
private int endTime;
private int appliedCloud;
List<Integer> directPriors;

public Task(int id, int[] executionTime,
    int canStartTime, List<Integer>
    directPriors) {
        super();
        this.id = id;
        this.executionTime = executionTime
            ;
        this.canStartTime = canStartTime;
        this.iniCanStartTime =
            canStartTime;
        this.endTime = Integer.MAX_VALUE;
        this.appliedCloud = -1;
        this.directPriors = directPriors;
}

public void reset() {
        this.canStartTime = this.
            iniCanStartTime;
        this.endTime = Integer.MAX_VALUE;
}

public int getId(){
        return id;
}

public int getEndTime() {
        return endTime;
}

public void setEndTime(int endTime) {
        this.endTime = endTime;
}

public int getAppliedCloud() {
```

```java
        return appliedCloud;
    }

    public void setAppliedCloud(int
        appliedCloud) {
            this.appliedCloud = appliedCloud;
    }

    public int getCanStartTime() {
            return canStartTime;
    }

    public void setCanStartTime(int
        canStartTime) {
            this.canStartTime = canStartTime;
    }

    public int[] getExecutionTime() {
            return executionTime;
    }

    public List<Integer> getDirectPriors() {
            return directPriors;
    }

}
```

C.3 PROGRAM 3: TASKGRAPH.JAVA

```java
import java.util.ArrayList;
import java.util.List;

public class TaskGraph {

    private List<Task> tasks = new ArrayList<Task>()
        ;
    private int cloudNum;
```

```java
    private int minExecutionTime = Integer.MAX_VALUE
        ;
    private int[] minAppliedServer;

    public TaskGraph(int cloudNum) {
        this.cloudNum = cloudNum;
}

    public void add(Task task){
        this.tasks.add(task);
}

    public void findOptimalSchedual(){
        minAppliedServer = new int[tasks.size()];
        bruteForce(0);

        System.out.println(minExecutionTime);
        for(int i=0; i<minAppliedServer.length-1;
            i++){
                System.out.print(minAppliedServer[
                    i] + ",_");
        }
        System.out.println(minAppliedServer[
            minAppliedServer.length-1]);
}

    private void permutate(int level, Task[]
        orderOfTasks){
            if(level >= orderOfTasks.length){
                int executionTime = execution(
                    orderOfTasks);
                if(executionTime <
                    minExecutionTime){
                        minExecutionTime =
                        executionTime;
                        for(int i=0; i<tasks.size
                        (); i++){
                            minAppliedServer[i
                            ] = tasks.get(i
                            ).
```

```
                              getAppliedCloud
                              ();
                }
        }
    }else{
            task:for(int  i=0; i<tasks.size();
                i++){
                    Task task = tasks.get(i);
                    int countOfPriors = 0;
                    List<Integer> priors =
                        task.getDirectPriors();
                    for(int  j=0; j<level; j++)
                        {
                            if(orderOfTasks[j]
                                == task){
                                    continue
                                        task;
                            }
                            if(priors.contains
                                (orderOfTasks[j
                                ].getId())){
                                    countOfPriors
                                        ++;
                            }
                    }
                    if(countOfPriors == priors
                        .size()){
                            orderOfTasks[level
                                ] = task;
                            permutate(level+1,
                                orderOfTasks);
                    }else{
                            continue task;
                    }
            }

    }
}

    private void bruteForce(int  level){
```

```
if ( level < tasks . size ()){
        for ( int i =0; i<cloudNum ; i++){
                tasks . get ( level ) .
                        setAppliedCloud ( i ) ;
                bruteForce ( level +1);
        }
} else {
        for ( int i =0; i<tasks . size () ; i++){
                tasks . get ( i ) . reset () ;
        }

        Task [] orderOfTasks = new Task [
                tasks . size () ];
        permutate (0 , orderOfTasks ) ;
    }
}

public int execution ( Task [] orderOfTasks ){
        int [] cloudAvailable = new int [cloudNum ];
        for ( int i =0; i<orderOfTasks . length ; i++){
                if ( orderOfTasks [ i ] . getCanStartTime
                        () == Integer .MAX_VALUE) {
                        orderOfTasks [ i ] .
                                setCanStartTime (0) ;
                }
                for ( int j =0; j<orderOfTasks [ i ] .
                        getDirectPriors () . size () ; j++){
                        if ( orderOfTasks [ i ] .
                                getCanStartTime () <
                                tasks . get ( orderOfTasks [
                                i ] . getDirectPriors () .
                                get ( j )) . getEndTime () ){
                                        orderOfTasks [ i ] .
                                                setCanStartTime
                                                ( tasks . get (
                                                orderOfTasks [ i
                                                ] .
                                                getDirectPriors
                                                () . get ( j )) .
                                                getEndTime ()) ;
```

```
                    }
               }
               if(orderOfTasks[i].getCanStartTime
                    () < cloudAvailable[
                    orderOfTasks[i].getAppliedCloud
                    ()]){
                        orderOfTasks[i].
                            setCanStartTime(
                            cloudAvailable[
                            orderOfTasks[i].
                            getAppliedCloud()]);
               }
               orderOfTasks[i].setEndTime(
                    orderOfTasks[i].getCanStartTime
                    () + orderOfTasks[i].
                    getExecutionTime()[orderOfTasks
                    [i].getAppliedCloud()]);
               cloudAvailable[orderOfTasks[i].
                    getAppliedCloud()] =
                    orderOfTasks[i].getEndTime();
          }

          int executionTime = 0;
          for(int i=0; i<tasks.size(); i++){
               if(executionTime < tasks.get(i).
                    getEndTime()){
                        executionTime = tasks.get(
                            i).getEndTime();
               }
          }
          return executionTime;
     }

     public int getMinExecutionTime() {
          return minExecutionTime;
     }

     public int[] getMinAppliedServer() {
          return minAppliedServer;
     }
}
```

```
public static void main(String[] args) {
    TaskGraph tg = new TaskGraph(3);
    int[][][] excutionTime =
        {
                        {3, 2, 5},
                        {5, 3, 4},
                        {7, 4, 5},
                        {10, 9, 12},
                        {6, 7, 8},
                        {4, 5, 9},
                        {3, 2, 4},
                        {8, 4, 5},
                        {6, 9, 12},
                        {13, 8, 5},
                        {12, 10, 11},
                        {2, 1, 3}
        };

    List<Integer> priors = new ArrayList<
        Integer>();
    Task task = new Task(0, excutionTime[0],
        0, priors);
    tg.tasks.add(task);

    priors = new ArrayList<Integer>();
    priors.add(0);
    task = new Task(1, excutionTime[1],
        Integer.MAX_VALUE, priors);
    tg.tasks.add(task);

    priors = new ArrayList<Integer>();
    priors.add(0);
    task = new Task(2, excutionTime[2],
        Integer.MAX_VALUE, priors);
    tg.tasks.add(task);

    priors = new ArrayList<Integer>();
    priors.add(1);
    priors.add(4);
```

```
task = new Task(3, excutionTime[3],
    Integer.MAX_VALUE, priors);
tg.tasks.add(task);

priors = new ArrayList<Integer>();
priors.add(1);
task = new Task(4, excutionTime[4],
    Integer.MAX_VALUE, priors);
tg.tasks.add(task);

priors = new ArrayList<Integer>();
priors.add(2);
priors.add(4);
task = new Task(5, excutionTime[5],
    Integer.MAX_VALUE, priors);
tg.tasks.add(task);

priors = new ArrayList<Integer>();
priors.add(0);
task = new Task(6, excutionTime[6],
    Integer.MAX_VALUE, priors);
tg.tasks.add(task);

priors = new ArrayList<Integer>();
priors.add(3);
priors.add(4);
task = new Task(7, excutionTime[7],
    Integer.MAX_VALUE, priors);
tg.tasks.add(task);

priors = new ArrayList<Integer>();
priors.add(5);
priors.add(7);
task = new Task(8, excutionTime[8],
    Integer.MAX_VALUE, priors);
tg.tasks.add(task);

priors = new ArrayList<Integer>();
priors.add(8);
priors.add(6);
```

```java
task = new Task(9, excutionTime[9],
    Integer.MAX_VALUE, priors);
tg.tasks.add(task);

priors = new ArrayList<Integer>();
task = new Task(10, excutionTime[10], 0,
    priors);
tg.tasks.add(task);

priors = new ArrayList<Integer>();
priors.add(10);
task = new Task(11, excutionTime[11],
    Integer.MAX_VALUE, priors);
tg.tasks.add(task);


long time1 = System.currentTimeMillis();
tg.findOptimalSchedual();
long time2 = System.currentTimeMillis();
System.out.println(time2 - time1);
```

References

[1] J. Wang, M. Qiu, B. Guo, and Z. Zong. Phase-reconfigurable shuffle optimization for Hadoop MapReduce. *IEEE Transactions on Cloud Computing*, PP(99):1, 2015.

[2] CISCO. Network virtualization-Access control design guide, 2008. url=http://www.cisco.com/c/en/us/td/docs/ solutions/Enterprise/Network_Virtualization/AccContr. html.

[3] K. Gai and S. Li. Towards cloud computing: A literature review on cloud computing and its development trends. In *The 4th International Conference on Multimedia Information Networking and Security*, pages 142–146, Nanjing, China, 2012. IEEE.

[4] K. Gai, M. Qiu, H. Zhao, L. Tao, and Z. Zong. Dynamic energy-aware cloudlet-based mobile cloud computing model for green computing. *JNCA*, 59:46–54, 2016.

[5] K. Gai, M. Qiu, and X. Sun. A survey on FinTech. *Journal of Network and Computer Applications*, PP:1, 2017.

[6] M. Qiu, M. Zhong, J. Li, K. Gai, and Z. Zong. Phase-change memory optimization for green cloud with genetic algorithm. *IEEE Transactions on Computers*, 64(12):3528–3540, 2015.

[7] K. Gai, M. Qiu, and H. Zhao. Cost-aware multimedia data allocation for heterogeneous memory using genetic algorithm in cloud computing. *IEEE Transactions on Cloud Computing*, PP(99):1–11, 2016.

[8] W. Dai, H. Chen, W. Wang, and X. Chen. RMORM: A framework of multi-objective optimization resource management in clouds. In *IEEE 9th World Congress on Services*, pages 488–494, Santa Clara, CA, USA, 2013.

[9] K. Gai and A. Steenkamp. A feasibility study of Platform-as-a-Service using cloud computing for a global service organization. *Journal of Information System Applied Research*, 7:28–42, 2014.

[10] K. Gai. A review of leveraging private cloud computing in financial service institutions: Value propositions and current performances. *Int'l J. of Computer Applications*, 95(3):40–44, 2014.

[11] K. Gai, M. Qiu, H. Zhao, and M. Liu. Energy-aware optimal task assignment for mobile heterogeneous embedded systems in cloud computing. In *2016 IEEE 3rd International Conference on Cyber Security and Cloud Computing (CSCloud)*, pages 198–203, Beijing, China, 2016. IEEE.

[12] K. Gai, M. Qiu, X. Sun, and H. Zhao. Security and privacy issues: A survey on FinTech. In *International Conference on Smart Computing and Communication*, pages 236–247, Shenzhen, China, 2016. Springer.

[13] C. Alcaraz and J. Aguado. MonPaaS: An adaptive monitoring Platform-as-a-Service for cloud computing infrastructures and services. *IEEE Transactions on Services Computing*, 8(1):65–78, 2015.

[14] J. Espadas, A. Molina, G. Jiménez, M. Molina, R. Ramírez, and D. Concha. A tenant-based resource allocation model for scaling Software-as-a-Service applications over cloud computing infrastructures. *Future Generation Computer Systems*, 29(1):273–286, 2013.

[15] D. Linthicum. *Cloud Computing and SOA Convergence in Your Enterprise: A Step-by-Step Guide*. Pearson Education, 2009.

[16] S. Brandt, E. Miller, D. Long, and L. Xue. Efficient metadata management in large distributed storage systems. In *2013 IEEE 10th International Conference on Mobile Ad-Hoc and Sensor Systems*, pages 290–290, Hangzhou, China, 2003.

[17] M. Lin, L. Xu, L. Yang, X. Qin, N. Zheng, Z. Wu, and M. Qiu. Static security optimization for real-time systems. *IEEE Transactions on Industrial Informatics*, 5(1):22–37, 2009.

[18] A. Steenkamp, A. Alawdah, O. Almasri, K. Gai, N. Khattab, C. Swaby, and R. Abaas. Teaching case enterprise architecture specification case study. *Journal of Information Systems Education*, 24(2):105, 2013.

[19] K. Gai, M. Qiu, H. Zhao, and W. Dai. Anti-counterfeit schema using monte carlo simulation for e-commerce in cloud systems. In *The 2nd IEEE International Conference on Cyber Security and Cloud Computing*, pages 74–79, New York, USA, 2015. IEEE.

[20] Y. Li, K. Gai, L. Qiu, M. Qiu, and H. Zhao. Intelligent cryptography approach for secure distributed big data storage in cloud computing. *Information Sciences*, 387:103–115, 2017.

[21] K. Gai, M. Qiu, and H. Zhao. Security-aware efficient mass distributed storage approach for cloud systems in big data. In *The 2nd IEEE International Conference on Big Data Security on Cloud*, pages 140–145, New York, USA, 2016.

[22] J. Wang, B. Guo, M. Qiu, and Z. Ming. Design and optimization of traffic balance broker for cloud-based telehealth platform. In *2013 IEEE/ACM 6th International Conference on Utility and Cloud Computing*, pages 147–154. IEEE, 2013.

[23] M. Qiu, W. Dai, and Keke K. Gai. *Mobile Applications Development with Android: Technologies and Algorithms*. CRC Press, 2016.

[24] K. Gai, M. Qiu, H. Zhao, and X. Sun. Resource management in sustainable cyber-physical systems using heterogeneous cloud computing. *IEEE Transactions on Sustainable Computing*, PP(99):1, 2017.

[25] K. Gai, Z. Du, M. Qiu, and H. Zhao. Efficiency-aware workload optimizations of heterogeneous cloud computing for capacity planning in financial industry. In *The 2nd IEEE International Conference on Cyber Security and Cloud Computing*, pages 1–6, New York, USA, 2015. IEEE.

[26] J. Yu, Y. Zhu, L. Xia, M. Qiu, Y. Fu, and G. Rong. Grounding high efficiency cloud computing architecture: HW-SW co-design and implementation of a stand-alone web server on FPGA. In

2011 Fourth International Conference on the Applications of Digital Information and Web Technologies, pages 124–129, Stevens Point, WI, 2011.

[27] L. Chen, Y. Duan, M. Qiu, J. Xiong, and K. Gai. Adaptive resource allocation optimization in heterogeneous mobile cloud systems. In *The 2nd IEEE International Conference on Cyber Security and Cloud Computing*, pages 19–24, New York, USA, 2015. IEEE.

[28] H. Zhao, M. Qiu, K. Gai, J. Li, and X. He. Maintainable mobile model using pre-cache technology for high performance android system. In *The 2nd IEEE International Conference on Cyber Security and Cloud Computing*, pages 175–180, New York, USA, 2015. IEEE.

[29] H. Zhao, M. Chen, M. Qiu, K. Gai, and M. Liu. A novel pre-cache schema for high performance Android system. *Future Generation Computer Systems*, 56:766–772, 2016.

[30] L. Qiu, K. Gai, and M. Qiu. Optimal big data sharing approach for tele-health in cloud computing. In *The IEEE International Conference on Smart Cloud 2016*, pages 184–189, New York, USA, 2016. IEEE.

[31] H. Zhang, G. Jiang, K. Yoshihira, H. Chen, and A. Saxena. Intelligent workload factoring for a hybrid cloud computing model. In *2009 World Conference on Services-I*, pages 701–708, Los Angeles, CA, USA, 2009. IEEE.

[32] H. Dai, Q. Li, M. Qiu, Z. Yu, and Z. Jia. A cloud trust authority framework for mobile enterprise information system. In *2014 IEEE 8th International Symposium on Service Oriented System Engineering*, pages 496–501, Oxford, UK, 2014.

[33] R. Dodda, C. Smith, and A. van Moorsel. An architecture for cross-cloud system management. In *Contemporary Computing*, pages 556–567. Springer, 2009.

[34] F. Hu, M. Qiu, J. Li, T. Grant, D. Taylor, and S. McCaleb et al. A review on cloud computing: Design challenges in architecture and security. *J. of Computing and Info. Tech.*, 19(1):25–55, 2011.

[35] A. Beloglazov and R. Buyya. Managing overloaded hosts for dynamic consolidation of virtual machines in cloud data centers under quality of service constraints. *IEEE Transactions on Parallel and Distributed Systems*, 24(7):1366–1379, 2013.

[36] A. Beloglazov and R. Buyya. OpenStack neat: A framework for dynamic and energy-efficient consolidation of virtual machines in OpenStack clouds. *Concurrency and Computation: Practice and Experience*, 27(5):1310–1333, 2015.

[37] M. Qiu, Z. Ming, J. Li, , S. Liu, B. Wang, and Z. Lu. Three-phase time-aware energy minimization with DVFS and unrolling for chip multiprocessors. *Journal of System Architecture*, 58(10):439–445, 2012.

[38] Y. Chen, M. Alghamdi, X. Qiu, J. Zhang, M. Jiang, and M. Qiu. TERN: A self-adjusting thermal model for dynamic resource provisioning in data centers. In *The 17th IEEE International Conference on High Performance Computing and Communications*, pages 479–490, New York, USA, 2015. IEEE.

[39] M. Qiu, Z. Chen, J. Niu, G. Quan, X. Qin, and L. Yang. Data allocation for hybrid memory with genetic algorithm. *IEEE Transactions on Emerging Topics in Computing*, pp:1–11, 2015.

[40] K. Gai, M. Qiu, X. Sun, and H. Zhao. Smart data deduplication for telehealth systems in heterogeneous cloud computing. *Journal of Communications and Information Networks*, 1(4):93–104, 2016.

[41] M. Qiu, L. Yang, Z. Shao, and E. Sha. Dynamic and leakage energy minimization with soft real-time loop scheduling and voltage assignment. *IEEE Transations on Very Large Scale Integration System*, 18(3):501–504, 2010.

[42] M. Qiu, E. Sha, M. Liu, M. Lin, S. Hua, and L. Yang. Energy minimization with loop fusion and multi-functional-unit scheduling for multidimensional DSP. *J. of Parallel and Distributed Computing*, 68(4):443–455, 2008.

[43] J. Li, M. Qiu, Z. Ming, G. Quan, X. Qin, and Z. Gu. Online optimization for scheduling preemptable tasks on IaaS cloud systems.

Journal of Parallel and Distributed Computing, 72(5):666–677, 2012.

[44] W. Tian, Y. Zhao, M. Xu, Y. Zhong, and X. Sun. A toolkit for modeling and simulation of real-time virtual machine allocation in a cloud data center. *IEEE Transactions on Automation Science and Engineering*, 12(1):153–161, 2015.

[45] K. Gai, M. Qiu, and H. Hassan. Secure cyber incident analytics framework using Monte Carlo simulations for financial cybersecurity insurance in cloud computing. *Concurrency and Computation: Practice and Experience*, PP(99):1, 2016.

[46] K. Gai, M. Qiu, and S. Elnagdy. A novel secure big data cyber incident analytics framework for cloud-based cybersecurity insurance. In *The 2nd IEEE International Conference on Big Data Security on Cloud*, pages 171–176, New York, USA, 2016.

[47] H. Jean-Baptiste, M. Qiu, K. Gai, and L. Tao. Meta meta-analytics for risk forecast using big data meta-regression in financial industry. In *The 2nd IEEE International Conference on Cyber Security and Cloud Computing*, pages 272–277, New York, USA, 2015. IEEE.

[48] X. Yu, T. Pei, K. Gai, and L. Guo. Analysis on urban collective call behavior to earthquake. In *The IEEE Int'l Symp. on Big Data Security on Cloud*, pages 1302–1307, New York, USA, 2015. IEEE.

[49] H. Liang and K. Gai. Internet-based anti-counterfeiting pattern with using big data in china. In *The IEEE International Symposium on Big Data Security on Cloud*, pages 1387–1392, New York, USA, 2015. IEEE.

[50] K. Gai, M. Qiu, L. Chen, and M. Liu. Electronic health record error prevention approach using ontology in big data. In *17th IEEE International Conference on High Performance Computing and Communications*, pages 752–757, New York, USA, 2015.

[51] Y. Li, K. Gai, M. Qiu, W. Dai, and M. Liu. Adaptive human detection approach using FPGA-based parallel architecture in reconfigurable hardware. *Concurrency and Computation: Practice and Experience*, PP(99):1, 2016.

[52] H. Yin, K. Gai, and Z. Wang. A classification algorithm based on ensemble feature selections for imbalanced-class dataset. In *The 2nd IEEE International Conference on High Performance and Smart Computing*, pages 245–249, New York, USA, 2016.

[53] H. Yin and K. Gai. An empirical study on preprocessing high-dimensional class-imbalanced data for classification. In *The IEEE International Symposium on Big Data Security on Cloud*, pages 1314–1319, New York, USA, 2015.

[54] Internetlivestats.com. Twitter usage statistics, 2016. url=http://www.internetlivestats.com/twitter-statistics/.

[55] K. Gai, M. Qiu, and S. Elnagdy. Security-aware information classifications using supervised learning for cloud-based cyber risk management in financial big data. In *The 2nd IEEE International Conference on Big Data Security on Cloud*, pages 197–202, New York, USA, 2016. IEEE.

[56] K. Thakur, M. Qiu, K. Gai., and M. Ali. An investigation on cyber security threats and security models. In *The 2nd IEEE International Conference on Cyber Security and Cloud Computing*, pages 307–311, New York, USA, 2015. IEEE.

[57] K. Gai, M. Qiu, B. Thuraisingham, and L. Tao. Proactive attribute-based secure data schema for mobile cloud in financial industry. In *The IEEE International Symposium on BigDataSecurity*, pages 1332–1337, New York, USA, 2015.

[58] M. Qiu, K. Gai, B. Thuraisingham, L. Tao, and H. Zhao. Proactive user-centric secure data scheme using attribute-based semantic access controls for mobile clouds in financial industry. *Future Generation Computer Systems*, PP:1, 2016.

[59] K. Gai, M. Qiu, and H. Hassan. Secure cyber incident analytics framework using Monte Carlo simulations for financial cybersecurity insurance in cloud computing. *Concurrency and Computation: Practice and Experience*, PP(99):1–15, 2016.

[60] K. Gai, M. Qiu, Z. Ming, H. Zhao, and L. Qiu. Spoofing-jamming attack strategy using optimal power distributions in wireless smart grid networks. *IEEE Transactions on Smart Grid*, 99(PP):1, 2017.

[61] K. Gai, M. Qiu, H. Zhao, and W. Dai. Privacy-preserving adaptive multi-channel communications under timing constraints. In *The IEEE International Conference on Smart Cloud 2016*, page 1, New York, USA, 2016. IEEE.

[62] K. Gai, M. Qiu, H. Zhao, and J. Xiong. Privacy-aware adaptive data encryption strategy of big data in cloud computing. In *The 2nd IEEE International Conference of Scalable and Smart Cloud (SSC 2016)*, pages 273–278, Beijing, China, 2016. IEEE.

[63] Y. Li, W. Dai, Z. Ming, and M. Qiu. Privacy protection for preventing data over-collection in smart city. *IEEE Transactions on Computers*, 65:1339–1350, 2015.

[64] N. Cao, C. Wang, M. Li, K. Ren, and W. Lou. Privacy-preserving multi-keyword ranked search over encrypted cloud data. *IEEE Transactions on Parallel and Distributed Systems*, 25(1):222–233, 2014.

[65] Y. Li, K. Gai, Z. Ming, H. Zhao, and M. Qiu. Intercrossed access control for secure financial services on multimedia big data in cloud systems. *ACM Transactions on Multimedia Computing Communications and Applications*, 12(4s):67, 2016.

[66] L. Ma, L. Tao, Y. Zhong, and K. Gai. RuleSN: Research and application of social network access control model. In *IEEE International Conference on Intelligent Data and Security*, pages 418–423, New York, USA, 2016.

[67] L. Ma, L. Tao, K. Gai, and Y. Zhong. A novel social network access control model using logical authorization language in cloud computing. *Concurrency and Computation: Practice and Experience*, PP(99):1, 2016.

[68] M. Nitti, R. Girau, and L. Atzori. Trustworthiness management in the social Internet of things. *IEEE Transactions on Knowledge and Data Engineering*, 26(5):1253–1266, 2014.

[69] R. Chen, F. Bao, M. Chang, and J. Cho. Dynamic trust management for delay tolerant networks and its application to secure routing. *IEEE Transactions on Parallel and Distributed Systems*, 25(5):1200–1210, 2014.

[70] K. Gai, L. Qiu, M. Chen, H. Zhao, and M. Qiu. SA-EAST: security-aware efficient data transmission for ITS in mobile heterogeneous cloud computing. *ACM Transactions on Embedded Computing Systems*, 16(2):60, 2017.

[71] K. Gai, M. Qiu, S. Jayaraman, and L. Tao. Ontology-based knowledge representation for secure self-diagnosis in patient-centered telehealth with cloud systems. In *The 2nd IEEE International Conference on Cyber Security and Cloud Computing*, pages 98–103, New York, USA, 2015. IEEE.

[72] K. Ruan, J. Carthy, T. Kechadi, and I. Baggili. Cloud forensics definitions and critical criteria for cloud forensic capability: An overview of survey results. *Digital Investigation*, 10(1):34–43, 2013.

[73] K. Gai, M. Qiu, L. Tao, and Y. Zhu. Intrusion detection techniques for mobile cloud computing in heterogeneous 5G. *Security and Communication Networks*, 9(16):3049–3058, 2016.

[74] R. Deshmukh and K. Devadkar. Understanding DDoS attack & its effect in cloud environment. *Procedia Computer Science*, 49:202–210, 2015.

[75] Y. Zhang, A. Juels, M. Reiter, and T. Ristenpart. Cross-tenant side-channel attacks in PaaS clouds. In *Proceedings of the 2014 ACM SIGSAC Conference on Computer and Communications Security*, pages 990–1003, Scottsdale, Arizona, USA, 2014. ACM.

[76] D. Gullasch, E. Bangerter, and S. Krenn. Cache games-bringing access-based cache attacks on AES to practice. In *2011 IEEE Symposium on Security and Privacy*, pages 490–505, Berkeley, CA, USA, 2011. IEEE.

[77] Y. Yarom and K. Falkner. Flush+reload: A high resolution, low noise, l3 cache side-channel attack. In *23rd USENIX Security Symposium*, pages 719–732, San Diego, CA, USA, 2014.

[78] S. Elnagdy, M. Qiu, and K. Gai. Understanding taxonomy of cyber risks for cybersecurity insurance of financial industry in cloud computing. In *The 2nd IEEE International Conference of Scalable and Smart Cloud*, pages 295–300. IEEE, 2016.

[79] P. Viswanathan, V. Batra, and P. Vyas. Convenient use of push button mode of WPS (Wi-Fi protected setup) for provisioning wireless devices, November 17 2015. US Patent 9,191,771.

[80] W. Cope, L. Paczkowski, and W. Parsel. Multiple secure elements in mobile electronic device with near field communication capability, April 29 2014. US Patent 8,712,407.

[81] S. Bye and L. Paczkowski. Near field communication authentication and validation to access corporate data, July 23 2013. US Patent 8,494,576.

[82] Y. Wang and Y. Ha. FPGA-based 40.9-gbits/s masked AES with area optimization for storage area network. *IEEE Transactions on Circuits and Systems II: Express Briefs*, 60(1):36–40, 2013.

[83] Consumer Report. 3.1 million smart phones were stolen in 2013, nearly double the year before, 2014. url=http://pressroom.consumerreports.org/pressroom/2014/04/my-entry-1.html.

[84] M. Qiu, L. Zhang, Z. Ming, Z. Chen, X. Qin, and L. Yang. Security-aware optimization for ubiquitous computing systems with SEAT graph approach. *J. of Computer and Syst. Sci.*, 79(5):518–529, 2013.

[85] Z. Shao, C. Xue, Q. Zhuge, M. Qiu, B. Xiao, and E. Sha. Security protection and checking for embedded system integration against buffer overflow attacks via hardware/software. *IEEE Transactions on Computers*, 55(4):443–453, 2006.

[86] K. Gai, M. Qiu, B. Thuraisingham, and L. Tao. Proactive attribute-based secure data schema for mobile cloud in financial industry. In *The IEEE International Symposium on Big Data Security on Cloud; 17th IEEE International Conference on High Performance Computing and Communications*, pages 1332–1337, New York, USA, 2015.

[87] J. Song, V. Wong, V. Leung, and Y. Kawamoto. Secure routing with tamper resistant module for mobile ad hoc networks. *ACM SIGMOBILE Mobile Computing and Communications Review*, 7(3):48–49, 2003.

[88] K. Gai, M. Qiu, and H. Zhao. Privacy-preserving data encryption strategy for big data in mobile cloud computing. *IEEE Transactions on Big Data*, PP(99):1, 2017.

[89] O. Aldor and N. Solomon. Methods for protecting against cookie-poisoning attacks in networked-communication applications, May 8 2012. US Patent 8,176,539.

[90] A. Møller and M. Schwarz. Automated detection of client-state manipulation vulnerabilities. *ACM Transactions on Software Engineering and Methodology*, 23(4):29, 2014.

[91] N. Kumar and S. Sharma. Study of intrusion detection system for DDoS attacks in cloud computing. In *2013 Tenth International Conference on Wireless and Optical Communications Networks*, pages 1–5. IEEE, 2013.

[92] J. Long, B. Gardner, and J. Brown. *Google Hacking for Penetration Testers*, volume 2. Syngress, 2011.

[93] L. Tao, S. Golikov, K. Gai, and M. Qiu. A reusable software component for integrated syntax and semantic validation for services computing. In *The 9th International IEEE Symposium on Service-Oriented System Engineering*, pages 127–132, San Francisco Bay, USA, 2015. IEEE.

[94] G. Alipui, L. Tao, K. Gai, and N. Jiang. Reducing complexity of diagnostic message pattern specification and recognition on inbound data using semantic techniques. In *The 2nd IEEE International Conference of Scalable and Smart Cloud*, pages 267–272. IEEE, 2016.

[95] S. Jayaraman, L. Tao, K. Gai, and N. Jiang. Drug side effects data representation and full spectrum inferencing using knowledge graphs in intelligent telehealth. In *The 2nd IEEE International Conference of Scalable and Smart Cloud*, pages 289–294. IEEE, 2016.

[96] R. DeStefano, L. Tao, and K. Gai. Improving data governance in large organizations through ontology and linked data. In *The 2nd IEEE International Conference of Scalable and Smart Cloud*, pages 279–284. IEEE, 2016.

[97] C. Asamoah, L. Tao, K. Gai, and N. Jiang. Powering filtration process of cyber security ecosystem using knowledge graph. In *The 2nd IEEE International Conference of Scalable and Smart Cloud*, pages 240–246. IEEE, 2016.

[98] M. Sette, L. Tao, K. Gai, and N. Jiang. A semantic approach to intelligent and personal tutoring system. In *The 2nd IEEE International Conference of Scalable and Smart Cloud*, pages 261–266. IEEE, 2016.

[99] S. Elnagdy, M. Qiu, and K. Gai. Cyber incident classifications using ontology-based knowledge representation for cybersecurity insurance in financial industry. In *The 2nd IEEE International Conference of Scalable and Smart Cloud*, pages 301–306. IEEE, 2016.

[100] K. Gai and A. Steenkamp. Feasibility of a Platform-as-a-Service implementation using cloud computing for a global service organization. In *Proceedings of the Conference for Information Systems Applied Research*, volume 2167, page 1508, 2013.

[101] D. Austin, A. Barbir, C. Herris, and S. Garg. Web services architecture requirements, 2004. http://www.w3.org/TR/wsa-reqs/.

[102] E. Christensen, F. Curbera, G. Meredith, and S. Weerawarana. Web services description language (WSDL) 1.1, 2001. http://www.w3.org/TR/wsdl.

Index